Integrated Science

Volume 40

Editor-in-Chief
Nima Rezaei, Tehran University of Medical Sciences
Tehran, Iran

The **Integrated Science** Series aims to publish the most relevant and novel research in all areas of Formal Sciences, Physical and Chemical Sciences, Biological Sciences, Medical Sciences, and Social Sciences. We are especially focused on the research involving the integration of two of more academic fields offering an innovative view, which is one of the main focuses of Universal Scientific Education and Research Network (USERN), science without borders.

Integrated Science is committed to upholding the integrity of the scientific record and will follow the Committee on Publication Ethics (COPE) guidelines on how to deal with potential acts of misconduct and correcting the literature.

Mariafilomena Anzalone
Stefania Achella • Fiorella Battaglia
Anna Donise
Editors

Reconfiguring Human Autonomy

Conceptual Challenges and Ethical Implications in the Age of AI

Springer

Editors
Mariafilomena Anzalone
Dipartimento per l'Innovazione Umanistica, Scientifica e Sociale (DIUSS)
University of Basilicata
Potenza, Italy

Stefania Achella
Dipartimento di Scienze Filosofiche, Pedagogiche e Sociali
University of Chieti-Pescara
Chieti, Italy

Fiorella Battaglia
Dipartimento di Studi Umanistici
University of Salento
Lecce, Italy

Anna Donise
Dipartimento di Studi Umanistici
University of Naples Federico II
Naples, Italy

ISSN 2662-9461 ISSN 2662-947X (electronic)
Integrated Science
ISBN 978-3-032-14536-9 ISBN 978-3-032-14537-6 (eBook)
https://doi.org/10.1007/978-3-032-14537-6

This Springer imprint is published by the registered company Springer Nature Switzerland AG
The registered company address is: Gewerbestrasse 11, 6330 Cham, Switzerland

Competing Interests

This publication was developed within the framework of the European Union – NextGenerationEU, under the National Recovery and Resilience Plan (PNRR), Mission 4, Component 2, PRIN 2022 PNRR Research Project “DigitHuman: Digital Transition and Human Values”. CUP: F53D23010740001.

Preface

The integration of algorithms and artificial intelligence into diverse domains—including education, healthcare, caregiving, and warfare—poses significant challenges to longstanding philosophical frameworks traditionally employed to conceptualize, interpret, and characterize human beings. Among these, the concept of autonomy can be identified as a central and enduring principle within the tradition of modern philosophy. Against the backdrop of human autonomy, which is traditionally understood as rational self-determination grounded in a free and morally responsible will, the emergence of so-called "autonomous" machine agents—capable of learning, decision-making, and action—challenges established assumptions and calls for renewed theoretical and ethical reflection. In response to the urgency posed by these radical transformations, the concept of autonomy has once again become a central focus of contemporary philosophical and ethical debate, particularly within moral theory, where it plays a foundational role in articulating principles of responsibility, agency, and moral accountability. Over the centuries, the paradigm of personal autonomy has been articulated through various conceptual frameworks, including rational, moral, political, and relational interpretations. Although the concept of personal autonomy plays a central role in both moral and political philosophy, its precise definition and scope remain a matter of considerable debate. Several influential philosophers have developed and accepted analyses of autonomy closely aligned with the idea that an autonomous person is their own legislator—someone who governs themselves and determines their actions independently. This interpretation ranges from Immanuel Kant's foundational anchoring of autonomy in the moral rationality of the subject, through Harry J. Frankfurt's "self-government" perspective, to the feminist framework of relational autonomy advanced by Marina Oshana, Natalie Stoljar, and Catriona Mackenzie, which challenges the traditional notion of autonomy as simply being one's own legislator. Autonomy has consistently played a pivotal role in shaping the conception of human identity. Concurrently, the concept has gradually expanded, leading to the attribution of autonomy to increasingly broader and more diverse categories of subjects. However, in contemporary discourse, the attribute of "autonomous" is increasingly applied to objects as well, particularly in the context of advanced technologies. Autonomous machines and artificial intelligence systems—capable of learning, making decisions, and acting without direct human control—exemplify this shift. This extension of autonomy beyond human agents raises complex theoretical and ethical questions about the nature of agency, responsibility, and moral accountability in relation to non-human actors. With the design and construction of "autonomous" machines—systems to which we delegate tasks traditionally performed by humans, often involving moral responsibility—the notion that autonomy is an exclusive human attribute increasingly comes into question. Through the integration of artificial intelligence, machine learning, and advanced decision-making capabilities, these machines replicate human autonomous action in a manner that is unprecedented both in scope and complexity. This technological evolution raises profound ethical challenges, particularly regarding accountability, responsibility, and the delegation of moral agency to non-human actors. Philosophically, it calls into question longstanding assumptions about the uniqueness of human autonomy, prompting a re-examination of what it means to be an autonomous agent in a world increasingly shared with intelligent machines. The blurring boundaries between human and machine agency demand new theoretical frameworks that can address issues such as machine moral status, the limits of algorithmic decision-making, and the conditions under which autonomy can be ascribed beyond the human subject. We are confronted with a challenge whose full contours remain yet to be defined. In our view, addressing this challenge requires focusing on the connection between human vulnerability and the fundamentally relational nature of human autonomy—an aspect emphasized, for example, in the concept of empathic relationship.

This volume is driven by two central aims: to rethink human autonomy considering the rise of "machine" autonomies and, simultaneously, to critically examine the descriptive and normative boundaries attributed to these forms of autonomy. The idea is neither to reinforce yet another dichotomy of "us" versus "the machines" nor to endorse optimistic narratives that portray AI as a means for humanity's moral enhancement. Instead, this volume calls for an inversion of perspective—inviting readers to explore how debates surrounding the autonomy of artificial systems can "refract" and illuminate the limits, possibilities, and transformative potential inherent in human autonomy, both at the individual and collective levels. To this end, in contrast to oppositional approaches that regard machine autonomy as a threat to human autonomy, this book adopts a co-evolutionary and dialogical approach. This text explores how new forms of autonomy in artificial and technological contexts contribute to a broader, more pluralistic, and contextualized understanding of human autonomy. Rather than viewing autonomy as a fixed, monolithic notion, the text considers it a historically contingent concept subject to ongoing refinement. The challenge before us is to address these theoretical, ethical, and political issues by critically questioning the traditional approach. Rather than evaluating the "machinicity" of artificial autonomy solely through human-centric parameters, we investigate the evolving forms of autonomy with the aim of fundamentally reconsidering the concept of human autonomy itself.

Articulated into three sections, the volume navigates through conceptual genealogy, ethical considerations, and practical applications.

In the first part, "What Is Autonomy in the Age of AI?", contributions undertake the essential intellectual task of disentangling the idea of "autonomy" from its exclusive human-centric ties. In this section, the question of autonomy is also addressed from a historical-philosophical perspective, questioning its foundations and transformations in the light of comparisons with artificial intelligence. The contributions analyse the aporias of the modern concept of autonomy, explore feminist and relational perspectives, focus on the limits of traditional categories in describing the agency of machines, and propose a critical analysis of the mimetic and perception-based models that shape our relationship with artificial agents. Starting from Mariafilomena Anzalone's account of the imitative devices that inform AI (Chap. 1), through Stefania Achella's feminist reinterpretation (Chap. 2), to the delimitation of the criteria of agency and moral *patiency* in Paul Formosa, Inês Hipólito, and Thomas Montefiore's contribution (Chap. 3), this section shows how autonomy, far from being a unitary concept, is articulated in a plurality of dimensions: cognitive, relational, normative, perceptual, and projective. In this case, AI acts as a "litmus test", forcing us to distinguish and circumscribe what, in our tradition, has been called autonomy—freedom, responsibility, self-legislation—from what machines actually achieve—learning, adaptation, and action without supervision.

The second part, "Human and Machinic Autonomy. Ethical Implications", focuses on the ethical and social repercussions of the interaction between human agents and autonomous machine systems. This interaction is investigated with particular attention to specific autonomous technological systems such as AVs, personal avatars, autonomous weapons, and recommendation systems. By applying theoretical perspectives such as that of relational autonomy or reflecting on the conditions and specificities of personal and moral autonomy, the contributions of this section concretely explore the transformative effects of such interaction. Hugo Cossette-Lefebvre and Natalie Stoljar show how *recommender systems* shape—and sometimes distort—users' preferences, reproducing social biases (Chap. 4); Mario De Caro reopens the question of the moral status of strong AI (Chap. 5), whereas Fiorella Battaglia's contribution focuses on how the concept of personal autonomy changes when it is extended to artificial intelligence systems, with a specific focus on the domain of defence (Chap. 6); Fabio Fossa criticizes techno-solutionist views on human autonomy "enhanced" by autonomous vehicles (Chap. 7); Paula Sweeney, finally, investigates how personal avatars and synthetic agents challenge the identity continuity of the subject and the ownership of actions (Chap. 8). As a result, the essays in this section also draw a map of the new ethical-political challenges that emerge from the attribution of autonomy to artificial devices: from widespread responsibility and algorithmic manipulation to the displacement of action and new configurations of power.

The third part, "Autonomy and AI in Medicine and Health Care. Challenges, Problems, and Opportunities", explores one of the most delicate and significant fields of AI application: the health sector. Here, the focus falls on the key areas of health and care, where the value of autonomy meets the fragility of bodily vulnerability. In this context, the introduction of AI is generating enormous expectations, both from the point of view of medical and pharmacological research and for clinical practice and the management of health activities. In this context, the question of autonomy, as the essays in this section aptly show, must be addressed from several points of view—that of doctors, that of patients, and that of their relationship. Special focus is placed on the impact of predictive systems, diagnostic chatbots, and choice-supporting algorithmic tools on the decision-making autonomy of the patient and of the doctor, which redefines the terms and quality of their relationship. The authors' analytical viewpoints—which encompass the ethics of care, narrative medicine, culturally attuned healthcare, and biopolitical theory—illustrate that the concept of autonomy has become a crucial intersection of epistemological, cultural, and political issues in contemporary discourse. Miranda Boldrini argues in favour of a relational view of autonomy that can enrich the ethics of AI in the health sector, focusing attention on the issues of trust, responsibility, and care (Chap. 9). While assessing risks and opportunities deriving from the impact of AI in health decision-making, Andrea Berber and Jelena Mijić identify the interest and well-being of the patient as the reference value apt to balance patient autonomy and that of the doctor (Chap. 10). Deivide Garcia da Silva Oliveira elaborates and presents a philosophical model for the doctor-patient relationship in the age of AI inspired by Feyerabend's pluralism (Chap. 11). Chantal Marazia, Vasilija Rolfes, and Fabio De Sio focus on the concept of "algorithmic humility" to promote a relational and reflective approach to health care, in which patient autonomy is established as a dynamic and socially rooted process (Chap. 12). Finally, Chetan Shirvankar asks how to promote a human-centred care system, in the light of the growing datafication of the doctor-patient relationship, and the risks of epistemic injustice introduced by the anthropomorphization of machines in the clinical field (Chap. 13).

As seen from the individual contributions, the incorporation of AI in the most varied contexts of our private and public life already widely shows its enormous transformative potential. When systems are transparent and interactive, they can strengthen the perception of control and individual choice; in contrast, algorithmic opacity can precisely undermine the element of trust that supports autonomy. For this reason, the lived experiences of not only students, teachers, patients, and health professionals but also, more simply, all the various people who interact with AI for different reasons are crucial: determining when and why AI emancipates or makes people dependent means redefining autonomy on the basis of concrete relationships rather than abstract normative statements. It is precisely in this framework that the paradigm of *relational autonomy*, developed by feminist theory, can play a key role and is recalled in many of the contributions in this volume: overcoming an atomistic and rationalist approach requires recognizing the structural links and inequalities that AI risks amplifying. A path is therefore outlined to counter damage to the most vulnerable subjects and de-emphasize gender biases and algorithmic disparities. Moreover, the volume highlights the need to carefully circumscribe the analogy between human autonomy and machine autonomy. While AI systems can replicate some functions of autonomous decision and action, these functions do not erase the ontological and moral distance between a human agent—capable of self-reflection—and a system, however complex, still lacking that internal evaluative dimension. Even in the military, for example, delegating life and death decisions to autonomous systems is not equivalent to transferring moral autonomy, but it definitely introduces fractional and controversial responsibilities. In the health and training field, the tension between responsibility and human autonomy is particularly strong. Systems such as recommenders or virtual avatars redefine the relationship between subject and knowledge as well as between representation and action. However, their effectiveness in increasing capacity for action and accessibility to information is accompanied by the risk of addiction, loss of authority, and agency crises. The figure of the doctor, for example, is not replaced by an algorithm but is certainly transformed by it, while the empowerment of the patient can oscillate between information consumption and passive delegation.

This highlights a fundamental regulatory issue. To support or rebuild autonomy in an algorithmically driven world, AI systems must be transparent, culturally inclusive, and designed on the basis of ethical criteria. This entails a standpoint well beyond reductionist or techno-solutionist approaches. It also implies acknowledging the fact that autonomy is not a static attribute, typical of an isolated individual, but a dynamic, distributed process, negotiated in contexts of care, education, and technical-social interaction.

This collection of contributions therefore converges on one underlying claim: human autonomy and machine autonomy should not be understood in terms of mere conflict or substitution but rather in their complex interrelation. Intelligent are neither simple neutral "tools" nor independent subjects; they reorient social practices, forms of interaction, and decision-making processes, thus redefining the very conditions of the possibility of autonomous action. Symmetrically, our moral categories—responsibility, accountability, agency—condition the design, regulation, and use of AI.

This volume finally invites us to overcome a dichotomously binary vision: autonomy is not a fixed attribute to be defended or delegated but rather a dynamic field of negotiation in which the recognition of subjects, the distribution of skills, and the construction of inclusive future scenarios play out. For this reason, we think that reconfiguring autonomy means, ultimately, reconfiguring what is human today. This amounts to accepting that our practical identity no longer coincides with an ideal of rational and individual self-sufficiency but is articulated and structured in the social bonds, in the technical environments, and in the socio-algorithmic networks that we inhabit.

The complex nature of this process is reflected in the essays collected in this book with their wide range of analyses and proposals, with their primary objective being the critical, not naive, exploration of a key question. Moving between moral philosophy, gender studies, phenomenology, the philosophy of technology, medical humanities, and ethical engineering, they show how the challenge of AI can become an opportunity to expand, rather than narrow, the horizons of individual and collective autonomy.

With this volume, we hope to contribute to a critical discussion that does not yield naive enthusiasm or forms of apocalyptic Luddism, which are rather common in contemporary debate. This discussion should allow the reader to see in artificial intelligence not only a technical tool or an entity "other" from us but also a mirror—sometimes faithful, sometimes deforming—of our most deeply rooted categories of understanding and evaluation of reality. Precisely in this ambivalent duplicity of AI lies its philosophical richness. Here lies the opportunity to walk past AI as a mere object, the use of which should be regulated, and embrace the investigation of who we are and who we want to be in the world we are helping to build.

Potenza, Italy Mariafilomena Anzalone
Chieti, Italy Stefania Achella
Lecce, Italy Fiorella Battaglia
Naples, Italy Anna Donise

Acknowledgements

We are grateful to Tessa Marzotto Caotorta for her essential support in preparing this volume, including editing, formatting, reference verification, and coordinating communication with contributors.

Contents

Part I What Is Autonomy in the Age of AI?

1 Autonomous Like Us? Machinic Copies and Human Models in the Age of AI 3
Mariafilomena Anzalone

2 Relational Autonomy. Rethinking AI Ethics in the Light of Feminism 17
Stefania Achella

3 AI and the Relationship Between Agency, Autonomy, and Moral Patiency 35
Paul Formosa, Inês Hipólito, and Thomas Montefiore

Part II Human and Machinic Autonomy. Ethical Implications

4 Automated Gatekeepers: How Recommender Systems Shape and Constrain Autonomy 53
Hugo Cossette-Lefebvre and Natalie Stoljar

5 The Promises and Perils of AI: Navigating Humanity's Most Transformative Technology 69
Mario De Caro

6 Personal Autonomy and Autonomous Systems: Sameness and Difference 81
Fiorella Battaglia

7 Autonomy-Enhancing Technologies? Automated Vehicles and Autonomous Mobility 95
Fabio Fossa

8 Personal Avatars and Personal Autonomy 109
Paula Sweeney

Part III Autonomy and AI in Medicine and Health Care. Challenges, Problems, and Opportunities

9 Beyond (Relational) Autonomy: AI Healthcare Through the Lens of Care Ethics . . . 123
Miranda Boldrini

10 Balancing Autonomies: The Impact of AI on Patient and Physician Decision-Making in Healthcare . . . 139
Andrea Berber and Jelena Mijić

11 AI in Clinical Practice: A Feyerabendian Pluralistic Model for Physician-Patient Relationships . . . 155
Deivide Garcia da Silva Oliveira

12 Big Data and the "Small X". Patient Autonomy in the Age of AI . . . 171
Chantal Marazia, Vasilija Rolfes, and Fabio De Sio

13 Humanizing Machines: Unpacking the Care Deficit in Anthropomorphized Medical AI . . . 185
Chetan Mahadev Shirvankar

Part I

What Is Autonomy in the Age of AI?

Autonomous Like Us? Machinic Copies and Human Models in the Age of AI

1

Mariafilomena Anzalone

Abstract

Due to the rapid expansion of digitization technological processes, and to the combination of the potential of AI with that of robotics, machine autonomy is perceived as an increasingly precise "copy" of human autonomy, which is bound, potentially, to equal if not to exceed the original "model". Underpinning this vision, which raises a whole series of concerns, including ethical ones, is the mimetic paradigm that structures the current construction of autonomous machines and feeds often simplistic analogizations with human autonomy. This chapter aims to explore this mimetic paradigm distinguishing two levels of analysis: a "perceptual-projective" level and an "effective" level. As to the first, I review the elements that facilitate machines being perceived as autonomous agents by human users and thereby prompt anthropomorphic projections; as to the second, I evaluate the technical architectures of these machines and the aspects of human autonomy that they concretely replicate. I finally argue that the autonomy of AI systems replicates aspects of human operational autonomy related to automatisms and, in this sense, achieves something more akin to the self-management of an organism than to the self-governance of a human will.

Keywords

Autonomy · Automatisms · Autonomic systems · Self-managing · Self-rule · Self-government

M. Anzalone (✉)
Dipartimento per l'Innovazione Umanistica, Scientifica e Sociale (DIUSS), University of Basilicata, Potenza, Italy
e-mail: mariafilomena.anzalone@unibas.it

M. Anzalone et al. (eds.), *Reconfiguring Human Autonomy*, Integrated Science 40, https://doi.org/10.1007/978-3-032-14537-6_1

1 Introduction

In the short story series *The Murderbot Diaries* [1], set in a dystopian future where the universe has become the target of human colonial expansion, SecUnit, a kind of cyborg guard designed to ensure the safety of the humans to whom it is assigned, manages to hack its control module. From that moment on, no longer bound to obey the orders of its masters, SecUnit can choose what to do. Instead of attending to its tasks, SecUnit devotes its time to its favorite activity: sitting quietly on the couch bingeing a collection of illegally downloaded TV shows, movies and soap operas. From a sophisticated "object" subservient to human purposes, the machine evolves, becoming a real subject, increasingly similar to human beings. Whereas its autonomy was previously expressed in independently and intelligently carrying out missions decided by others, it now shows in a radically different way. Walking past a purely operational and executive dimension, previously bound to the objectives of others, SecUnit's autonomy is now a self-determination, namely the ability to act on the basis of reasons and motivations that reflect desires, aspirations and objectives that belong to it and with which it can identify. Even, or perhaps above all, when these aspirations are the expression of its "addictions", such as the addiction to TV shows.

With its funny and ironic approach, this series of stories addresses the topic of human–machine interaction in the light of the underlying ambiguity that has characterized and still characterizes the production of autonomy-endowed machines. In short, one could in fact argue that we want such machines to be autonomous like us, but just to a certain extent [2, pp. 4–5]. They should be autonomous while carrying out the tasks for which we design and build them, preferably freeing us from the need to even control them. However, their autonomy must also be limited to a very precise perimeter: that outlined first of all by our objectives, beyond which their autonomy is experienced as a threat, if not a true danger, to human autonomy.

Literature, as well as cinematography, has largely explored dystopian and apocalyptic scenarios, in which technological artifacts (i.e., robots, technical systems, etc.) emancipate themselves from the role of obedient executors of the human will and oppose it, in some cases annihilating it. Today, however, with the developments of AI, this imagery, particularly vivid in Western culture [2, 3], seems to be closer to translating into reality. Owing to the rapid expansion of digitization technological processes, and to the combination of the potential of AI with that of robotics, systems—such as robots, drones, and AVs—are proliferating, that are capable of learning from experience, of adapting and responding to changes in the operational context, and of choosing and completing a course of action without the remote intervention of a human operator. These technological devices, to which we delegate tasks that, for human beings, in some cases, also involve bearing moral responsibilities, often appear as an increasingly precise "copy" of their initial "model". This fuels the fear that this sort of growing "humanization" of machine autonomy may have far more harmful outcomes than having to deal with a slacker cyborg such as SecUnit.

In this chapter, I will try to explore precisely the mimetic paradigms that structure the current construction of autonomous machines and, at the same time, more or less explicitly feed often simplistic and hasty analogizations with human autonomy. To this end, I distinguish two levels of analysis: at the "perceptual-projective" level, I review the elements that facilitate machines being perceived as autonomous agents by human users and thereby prompt anthropomorphic projections; at the second "effective" level, I evaluate the technical architectures of these machines and the aspects of human autonomy that they concretely replicate.

2 Mimetic Paradigms, Anthropomorphic Projections, and Normative Canons

Right from the start, with Alan Turing's "imitation game" [4],[1] the design and development of AI has been marked by an explicit mimetic paradigm. Since Turing's attempt to test whether a machine could mimic human intelligent behavior so effectively to be indistinguishable from it, machines have become able to imitate even the way in which human beings communicate with each other[2] and learn from others and the environment.[3] Moreover, the very architectures of AI systems are openly inspired by the way in which human beings "function" in biological and cerebral terms. One could mention here the pioneering *Perceptron* [9, 10], developed in 1957 by Frank Rosenblatt to simulate how the human brain self-organizes to learn information [11]. This produced the first prototype of an artificial neural network.[4] Rosenblatt's intuition first introduced the idea of a bottom-up approach. The same intuition is still currently the basis of machine learning and deep learning algorithms, whose node network structure is called, unsurprisingly, a deep neural network.

The integration of AI and the mechanical structure of robots has then achieved a physical interaction with the environment. It is increasingly clear, then, also from a

[1] On Turing's test and its impact on AI development, see, among others [5].

[2] The attempt to imitate human language and, in particular, conversational behaviours is essential to the early chatbots such as ELIZA [6]. Today, thanks to Large Language Models (LLM), the development of conversational agents such as GPT-4 is so advanced that they can hold conversations in a way that closely resembles what human beings do, even imitating forms of emotional connection that are typical of human dialogical relations. On this topic, see [7].

[3] In the field of machine learning, imitation is a key learning methodology. It allows the training of autonomous systems or robots through observation and the replication of human behaviour. The Imitation Learning approach, which is inspired by how children learn via imitation, is especially useful in order to teach complex tasks (e.g., autonomous driving) without the need for explicit detailed coding. On future perspectives and challenges connected to this approach, see [8].

[4] The idea to imitate brain physiology in a device was first introduced in 1943 by McCulloch and Pitts [12]. An account on how artificial neural networks have developed in relation to the theories of self-organization is provided by Pasquinelli [13, Chap. 6].

perceptual perspective, that robotics is no longer limited to reproducing the sensory-motor functioning of some parts of the human body (e.g. arms, hands, etc.), which is applied mainly to building machines intended for the industrial environment. Robotics is currently also able to produce anthropomorphic machines, capable, for example, of entering the sphere of human relationships, and sharing not only physical spaces (such as workplaces) but also social space with human beings [2, Chap. 1].

The whole field of social robotics has the specific but complex goal of building artificial social agents, that is to say, real social partners, capable of interacting with human beings in a way that is not only credible and effective, but also engaging and satisfying. To this end, robots must incorporate a series of skills ranging from high-level communication, to the expression and recognition of emotions, to the use of "natural" signals, such as gestures and gazes [14]. In addition, they must exhibit them in a socially coordinated way, reacting autonomously and appropriately to what their interlocutor does. In this context, a calibrated anthropomorphic design[5] also helps to support the obvious extension of mimesis to the social-emotional behaviour of human beings. This triggers the perception of robots as an otherness that, if not completely familiar, is in any case not alien to the context. Robots are seen as being endowed with autonomous mental states and emotions of their own, but that are easily decodable by the user. Thanks to the developments of AI, these robots can now learn from interacting with humans, imitating patterns of social behavior and customizing interactions [17]. This only amplifies the perception of dealing with something increasingly human.

With regard to social robotics, it is certainly clear that current technological design not only exploits the human tendency to anthropomorphize non-human entities, but also, in many ways, activates and reinforces it [18]. Even where the anthropomorphic component of a robotic body is missing, such as in the interaction with chatbots and virtual agents,[6] the same operating modes are used by AI systems to encourage the attribution of a series of human qualities. They are, in fact, designed to carry out their tasks (e.g. find information or solutions, assist individuals in some activity, etc.) by exhibiting a pleasant and engaging style of interaction, as similar as possible to what a human being would do.

The effective combination of task objectives and social objectives, that is to say, of factors that simulate cognitive abilities and human behavioral modes, increases the perception of the "humanity" of the machine. However, even with no in-depth computer knowledge, already at an intuitive level, it is clear that, for example, a

[5]The reader should take into account, first of all, the uncanny elements, which, according to the Uncanny Valley theory [15], ensue from a too close similarity between anthropomorphic machines and human beings. Based on said theory, more studies have been carried out on the mechanisms involved in assessing human similarity, by exploring the interaction with robots that are true copies of human beings, such as the geminoids [16].

[6]According to Wallach and Allen [19, p. 3] also chatbots and virtual agents, although they have no material or digital body, can be categorized within the realm of social robotics, as long as they are designed to interact with human beings and they have operational autonomy.

home alarm system and a virtual assistant, such as for instance the medical scribe Sunoh.ai,[7] achieve different forms of autonomy. Nevertheless the same term is used for both types of machines. It is also clear that Sunoh.ai's autonomy seems much closer to human autonomy.

Even by restricting assessment to how AI systems actually perform their tasks, the imitation of abilities considered specifically human is obvious. In regard to autonomy, then, the imitation is especially clear. AI systems are characterized by the ability to perceive, learn, judge and act in an adaptive and coherent way, without the need for constant human intervention.

Although the existence of technical objects defined as "autonomous" is not a novelty of the twenty-first century, it is undeniable that current developments in AI systems have allowed a considerable increase in the levels of independence from human supervision achieved by some devices, and a considerable extension of the areas in which their autonomy can be exercised, with a consequent expansion of the tasks that these devices can perform. AI systems are currently acting in contexts that, until now, have only featured human agents, and sometimes, they also perform functions that, for humans, involve bearing moral responsibilities. As a result, they are identified not only as "autonomous" but also as real "agents" or even as "moral agents".[8] One might want to discuss whether or not the now extensive use of the expression "moral agent" in reference to artificial or, in any case, non-human entities is fully legitimate [21]. Nevertheless, it is clear that in these cases, machine autonomy is perceived as an increasingly sophisticated and precise "copy" of human autonomy, bound, potentially, to equal if not to exceed the original "model".

The progressive improvement of the technical imitation of human autonomy, feeding the perception of an imminent fading out of the asymptotic features of this imitative process, is a source of much concern. At various levels of the discussion surrounding AI, concern is expressed in the public debate, in the philosophical debate, and in the set of guidelines, ethical recommendations and regulations related to AI and developed in recent years.[9] In general, a tangible fear emerges in all discussions: precisely when a machine replicates something very similar to human moral autonomy, the latter will be weakened, if not truly endangered by the growing autonomy of artificial intelligence systems [23–25]. Human autonomy then becomes a regulatory canon with respect to which: (a) the rate of autonomization and

[7] Sunoh.ai is an AI tool designed to transcribe and summarize the conversations between doctors and patients as a support in preparing clinical documentation. It is one of the services offered by the eClinicalWorks platform in the field of Health IT.

[8] A review of the debate surrounding artificial moral agents, understood as AI systems capable of recognizing the morally relevant aspects in a situation and take them into account in their decisions and actions, is provided by Misselhorn in [20]. On the only superficial similarities between human personal autonomy and the autonomy of AI technologies in the military defence sector, see, in this volume, Fiorella Battaglia's Chap. 6.

[9] A descriptive analysis of topics included in about 200 open source guidelines on AI is provided by Nicholas Kluge Corrêa, Camila Trindade Galvão, and colleagues in [22].

consequent "humanization" of machines occurs; (b) the magnitude of the risks and/or benefits that the autonomy of AI systems entails for human beings is measured in terms of decrease/increase.

3 Automated, Autonomous, Autonomic

In light of the widespread perception that the increasing autonomization of machines is moving towards their "humanization", it is appropriate to dive more deeply into the strictly technical field to try to actually understand what type of autonomy machines develop, in particular, those equipped with AI systems. In this regard, one should note that, in general, autonomy is referred to in a descriptive sense, functional to list what tasks the machine is able to perform without the remote intervention of a human operator [26, p. 56]. The type of tasks and the independence from the human being in performing them are the two general parameters based on which a machine is defined as autonomous. This obviously qualifies machine autonomy as essentially operational and at the same time as relative, namely relative to a limited field of action and to a defined environment.

Many devices, even very different from each other, can be identified as autonomous in this sense: a home alarm system that warns us of an attempted break-in as well as a self-driving vehicle. Both meet the "for-by-where" conditions mentioned by Guglielmo Tamburrini: they are autonomous *for* a specific task, carried out independently of human intervention *by* a given kind of agent, *where* a given context allows it [26, p. 58].

Devices equipped with AI systems fall within the wide range of technologies that meet these conditions, but they obviously constitute a particular subgroup, both for the variety and complexity of the tasks they perform and for the high levels of independence from humans that they can achieve. They represent an evolution of that process that, starting from the mechanization of production and/or operational processes, has gone toward their automation, to arrive today at the development of autonomous systems.

Automation is, in fact, a key element in understanding the specificity of AI-enabled systems. Based on automatization, it implies that a certain operation is made automatic. A device, which is based on predetermined instructions, operates and performs a certain task without a human being having to intervene. For instance, the thermostat responsible for regulating the temperature in the house, when the latter reaches the threshold value predefined by the user, automatically "reacts" by turning on the heating. An automated process or system therefore presupposes a series of operations sequenced into simple decisions, made automatic according to the general scheme "do X when Y is detected", and managed according to predefined and extremely detailed rules. As many scenarios as possible need to be foreseen and the automated system is instructed according to a top-down model that, in computer language, is formalized in algorithms.

The autonomy achieved by automation is therefore a form of independence from human beings, limited to the purely executive aspects of a process, and guaranteed by the fact that the system complies with predefined rules. The parameters, rules, objectives, and methods of implementation always depend on individuals who, on the basis of their knowledge, organise and formalise an informed and finalised decision-making process.

Compared with this type of autonomy, that of systems equipped with AI is significantly different. First, AI systems do not foresee the mere application of predefined models based on logic, or on knowledge encoded by human experts, since AI is able to abstract from experience, according to a bottom-up inferential scheme. Once trained on a number of examples and data—which may or may not have been arranged by humans[10]—AI systems are able to process new data and process adaptive responses to unforeseen scenarios. Once humans have set the objectives (for example, for an AV, starting from a place X, to arrive at place Y, in the shortest possible time and at the maximum speed allowed), the system decides how to achieve them and does so, clearly, without the sequence of actions having been previously prescribed in all its steps.[11]

While in the case of automation the system is designed to operate on the basis of human knowledge, AI systems now rely on their ability to learn, perceive, manage, and combine information of a different nature (images, sounds, data from other systems, etc.), continuously updating the representation of their state in relation to that of the surrounding environment and, therefore, also continuously updating the timing and methods of execution of their objectives. Self-monitoring and self-regulation allow these systems a degree of flexibility and adaptability that is unknown to classic automated systems, which react instead to predefined inputs to achieve the required results. But, above all, even if guided by a problem posed by the user, AI systems choose how to solve it without rigidly following preset procedures, but looking, with a contextually aware approach, for the best course of action.

The independence from human beings that such systems achieve is not limited, therefore, to the executive level, but ranges from the perceptual-cognitive level to the decision-making level. Precisely on this latter level, with a view to a more refined increase in their autonomous capacities, many of the technical challenges connected to such systems are played out. One clear challenge is connected to the objective of developing autonomic systems [27, 28], which represent a further and sophisticated evolution, compared with autonomous systems. While sharing with the latter the ability to self-monitor and self-regulate, autonomic systems are characterized by a distinctive capacity for self-management. Faced with failures or changes in external conditions, they are able to reconfigure and update their leading algorithms without the need for further input or human intervention from the outside. And, if, for example, the update is not beneficial, they are able to return to the previous configuration. Their autonomy is thus exercised not only in relation to the

[10] In general, within the framework of machine learning, AI is trained on data which is structured and monitored by human beings. When this is not the case, as within the framework of deep learning, relying on artificial neural networks the AI system autonomously learns how to categorize data.

[11] On this topic, see, in this volume, Fabio Fossa's Chap. 7.

external environment, but also in relation to their internal environment (i.e., the operating system, the database management software, etc.).

Systems of this kind, by virtue of their characteristics, can manage particularly complex IT systems such as contemporary ones, which integrate heterogeneous and interconnected environments. They can guarantee stability, efficiency and continuous performance optimization, as well as adaptive responses to unpredictable changes in the operating context so rapid that they are nearly automatic. Automatisms and automatizing processes, in fact, do not disappear at all within autonomic systems, whose distinctive feature is precisely being able to achieve an increasingly powerful automatization of their dynamic configuration [29, p. 17492]. Nor high-level general objectives and the rules preset by human operators to define the framework within which the system operates disappear. However, the fact that it is self-managing, self-correcting, and self-repairing, frees human beings from further coding and control tasks, and makes its autonomy seem particularly similar to that of humans.

This impression, although not completely unfounded, is, however, only partially correct. What inspires the approach of autonomic computing, in fact, is not so much a conscious human behavioral model, but a biological model, specifically that of the autonomic nervous system (ANS) [27].[12] Effectively ensuring the balance of internal conditions in the human body, ANS controls the smooth muscles, the heart muscle and the exocrine glands, regulating physiological processes essential for the survival of the body, such as heart rate and body temperature. Unlike the somatic nervous system (SNS), which controls voluntary movements, the ANS manages involuntary and automatic functions and does so without humans necessarily having to be aware of them or constantly dealing with them. Autonomic systems explicitly aim to reproduce precisely the relationship that human beings have with their ANS so that, while the latter works in the background, guaranteeing the regular functioning of their organs, they can do something else, focusing on operations of which they are aware.

However, as the developers of these systems point out, even in the explicit imitation of the ANS, autonomic systems have a fundamental difference: "Many of the decisions made by autonomic capabilities in the body are involuntary. In contrast, self-managing autonomic capabilities in computer systems perform tasks that IT professionals choose to delegate to the technology according to policies. Adaptable policy—rather than hard-coded procedure—determines the types of decisions and actions that autonomic capabilities perform" [27, p. 4].

Autonomic systems, therefore, are self-managed but are still governed by procedures and rules established upstream to ensure efficient operation, which is consistent with the safety and task objectives set by the developers.

[12] Senior vice president of IBM Research, Paul Horn presented the idea to develop autonomic systems at a National Academy of Engineers conference at Harvard University, stating that the inspiration for his idea came from how the ANS works [28, p. 41; 30].

4 Which Autonomy? Self-Managing, Self-Rule, Self-Government

In light of the previous brief survey of the type of autonomy developed by AI systems, one might wonder if and in what terms it constitutes a "copy" of human autonomy. This entails, however, a non-negligible difficulty. While machine autonomy, in the techno-scientific field, is defined quite unambiguously [31], in the philosophical field the conceptualizations of human autonomy are multiple and often contrasting.[13] Under the pressure of the demands of feminist and post-humanist philosophy,[14] even consolidated theoretical approaches—for example, Kant's theory of autonomy as a source of moral obligation, and as the one human quality that commands the moral respect of people—have undergone profound critical revisions. In particular, the limits and risks associated with an uncritical assumption of accounts of personal and moral autonomy based on the idea of rational and cognitive self-government have been exposed [33]. Moreover, at the same time, the relevance of the relational and affective components of human autonomy has come to the forefront [34]. This has most notably made clear, also in the context of AI ethics, the distorting effect that the surreptitious assumption of some visions of autonomy can have in identifying the ethical challenges connected to machinic autonomy today.[15]

There is no doubt that the very scenario shaped by the development of AI has further stimulated philosophical discussions and, in particular, ethical discussions, to refocus attention on the question of the nature of human autonomy: what is the human autonomy we want to protect from the growing machine autonomy or rather hope to strengthen thanks to it? Is it the moral autonomy of Kantian descent? Or is it personal autonomy that is exercised in relation to any dimension of human life, not just the moral aspect of it [35, pp. 34–47]? And under what conditions can an individual be considered autonomous? Are conditions related to rationality sufficient or should we also consider conditions related to relationality, such as the ability to give and ask for reasons [36]?

Two main theoretical approaches can be identified in the, certainly varied, answers to these questions emerged in the contemporary debate: that of procedural theories and that of substantial theories of autonomy [25, pp. 597–599]. The former,[16] although with noticeable differences, tend to identify formal procedures, that are neutral with respect to the content of personal decisions, on the basis of which to determine the necessary and sufficient conditions to be autonomous. In this view, subjects are autonomous to the extent that they possess certain skills and apply them in making their choices: self-control, rational thinking, freedom from systematic self-deception, ability to identify with their own desires, values, etc., and so on. Instead, substantial theories argue that procedural theories fail to address the problem of "oppressive socialization" [34, 38] and the related problem of adaptive

[13] A brief overview is provided by Christman in [32].

[14] On this topic, see, in this volume, Stefania Achella's Chap. 2.

[15] On this topic, see, in this volume, Miranda Boldrini's Chap. 9.

[16] Among these, one should mention, at least, Frankfurt's theory [37].

preferences [39], since they focus only on the formal characteristics through which the subject comes to choose an option, and not on the contents of the options available to them, which can potentially be the expression of conditions of oppression and/or vulnerability of various kinds. These theories then support an account of personal autonomy rooted in the reference to the content of norms or values considered valid and just, because, for example, they are not oppressive, but rather respectful of the dignity of all rational agents [39]. They also enhance the role of the relational components of autonomy both in the processes that preside over the deliberation and in those behind the development of the subject's idea of themselves.

Beyond the relevant differences in these two theoretical approaches, it is important to note that they both offer a range of "indicators" that, both taken individually and as a whole, allow us to converge on the fact that machines driven by AI systems do not replicate key aspects in the deployment of human autonomy. Although equipped with decision-making skills and some perceptual, cognitive and operational skills that are also present in autonomous subjects, machines have no desires or reasons of their own to reflect critically, or to identify with or to respond to others also on an emotional and sentimental level.[17] While choosing between different options without having to rigidly perform preset procedures, but rather identifying the best option, machines do not act freely according to their own rules, norms, and/or values. Their behavior—although inferred from the available data and not prescribed by a previous and detailed technical formalization of the problem to be solved—responds to goals, objectives and general rules that are not the expression of the machine's voluntary choice, but rather of one or more human choices.

This is particularly clear in the case of autonomic systems. They might well manage themselves, but they do so within the procedures and rules established by the developers with a view to making the system more efficient, consistently with their objectives. Their autonomy unfolds in terms of an operational and practical ability to plan activities, organize available resources, and efficiently perform assigned tasks. Moreover, it is rooted in an increasingly advanced form of automation. In the design and structuring of the machinic autonomy of AI systems, in fact, a key role is played precisely by automatisms. This type of autonomy is not only built from automated operations—such as machine learning—but it also aims, as in the case of autonomic systems, to make automatic a whole series of aspects related to their self-management—what humans used to manage—to allow these systems to take care of increasingly complex and articulated processes, responding adaptively to inputs from the outside.

This dynamic development emphasises the role of automatisms in view of increasing operational autonomy; one can read in it a significant analogy with a typically human mechanism, which is useful for the development of autonomous and intelligent behaviours, but is not exactly the same thing. What is at stake here is the mechanism of relief (*Entlastung*) which, according to Gehlen [40], is the essential adaptive strategy that has allowed human beings not only to survive, despite

[17] On the importance of reactive attitudes like blame or resentment for autonomy, see, in this volume, Fiorella Battaglia's Chap. 6.

their organic and biological deficiencies, but also to transform the environment into their world. According to Gehlen, since human beings lack highly specialized and rigid instincts allowing them to respond mechanically to stimuli and automatically adapt to a specific habitat, they establish fixed behavioral patterns that relieve them from continuously and repeatedly making all choices in response to environmental stimuli. This mechanism, which, for Gehlen, is at work at all levels of individual life,[18] featuring learning and habit as its essential components, allows the individual to economize energies, reserving them for more articulated and higher-level performance.

AI systems, through an increasing automation of their dynamic configuration, reproduce, albeit with technological solutions, this type of mechanism, which allows them to self-manage in the execution of increasingly complex tasks. However, their self-management does not coincide with a full form of self-government, as long as the procedures and rules that allow them to function are set by others.

One can therefore argue that machines driven by AI systems are self-managing but not self-governing in a fully human sense, where self-governance presupposes not only practical-operational independence, but also the ability to give themselves rules (self-rules), which are the expression of the will to adhere to one's own principles and values, and to do so, possibly, in opposition to any "external" government, namely to any pre-imposed system of rules. Instead, these systems are explicitly designed to respond as automatically as possible to a set of rules and purposes that do not belong to them and that, ultimately, are beyond their control.

5 Conclusion

The analysis of the mimetic paradigm underlying the creation of autonomous machines driven by AI has shown that the margin of comparison with human autonomy is rather limited and mostly concerns the exercise of some perceptual, cognitive, and decision-making skills in an operational-executive framework.

The autonomy of intelligent machines, even in their most advanced developments, is, in fact, deployed in terms of an adaptive self-management of highly complex systems, made possible by a quantitative and qualitative increase in automated operations. Specifically, the key role of automation processes also for AI systems contradicts the widespread perception that machine autonomy perfectly matches human autonomy. This is so not because automatisms are absent in the development of autonomous human behaviour. Gehlen made philosophically clear that the opposite is true. However, in the case of machine autonomy, automatisms are not only a tool to reach autonomy, but also the goal of its very development. In other words,

[18] At motor level, conditioned reflexes settle in; at perceptual level, relief can be observed in some fixations in the habits of seeing, listening, and touching; and at the level of communication and language, one is relieved from constantly referring to an object or to a perceptual experience by symbols and words which make allusions to them. Relief operates also at the level of thinking, through settled conceptual connections and habitual representations.

the goal of machine autonomy is just to make these machines capable of automatically carrying out increasingly complex tasks hitherto entrusted to human beings.

Moreover, even when automation is deployed in a sophisticated way, as in the case of AI systems, it is achieved in the repetitive and mechanical adherence to a rule [13, chap. 6]; autonomy, on the contrary, requires not only the ability to establish one's own rules, in which one identifies, and for which one responds to others, but also the ability to break rules given by others, as soon as they are considered oppressive or harmful to one's own dignity.

Granted that the humanization of the autonomy of a machine passes through the ability to disobey rules imposed by others—as in the case of SecUnit merrily slacking off on the sofa—then there is no doubt that this scenario, in which the machinic "copies" equal the human "model", is still a matter of science fiction.

Competing Interests This work was supported by the European Union – NextGenerationEU, under the National Recovery and Resilience Plan (PNRR), Mission 4, Component 2, PRIN 2022 PNRR Research Project "DigitHuman: Digital Transition and Human Values". CUP: F53D23010740001.

References

1. Wells M (2017–2025) The Murderbot Diaries series. Tor Books, New York
2. Dumouchel P, Damiano L (2017) Living with robots. Harvard University Press, Harvard
3. Battaglia F, Weidenfeld N (eds) (2014) Roboethics in film. Pisa University Press, Pisa
4. Turing A (1950) Computing machinery and intelligence. Mind 59(236):433–460
5. Oppy G, Dowe D (2021) The Turing test. In: Zalta EN (ed) The Stanford encyclopedia of philosophy (Winter 2021 edition). Stanford University. https://plato.stanford.edu/archives/win2021/entries/turing-test/
6. Weizenbaum J (1966) ELIZA—a computer program for the study of natural language communication between man and machine. Commun ACM 9:36–45. https://doi.org/10.1145/365153.365168
7. Shum HY, He XD, Li D (2018) From Eliza to XiaoIce: challenges and opportunities with social chatbots. Front Inf Technol Electron Eng 19:10–26. https://doi.org/10.1631/FITEE.1700826
8. Hussein A, Medhat Gaber M, Elyan E, Jayne C (2017) Imitation learning: A survey of learning methods. ACM Comput Surv 50(2):1–35. https://doi.org/10.1145/3054912
9. Rosenblatt F (1957) The Perceptron: A perceiving and recognizing automaton. Technical report project PARA. Cornell Aeronautical Laboratory, Buffalo. https://bpb-us-e2.wpmucdn.com/websites.umass.edu/dist/a/27637/files/2016/03/rosenblatt-1957.pdf. Accessed 3 July 2025
10. Rosenblatt F (1958) The Perceptron: A probabilistic model for information storage and organization in the brain. Psychol Rev 65(6):386–408. https://doi.org/10.1037/h0042519
11. Rosenblatt F (1961) Principles of neurodynamics: Perceptrons and the theory of brain mechanisms. Cornell Aeronautical Laboratory, Buffalo. https://gwern.net/doc/ai/nn/1962-rosenblatt-principlesofneurodynamics.pdf. Accessed 3 July 2025
12. McCulloch WS, Pitts W (1943) A logical calculus of the ideas immanent in nervous activity. Bull Math Biophys 5:115–133. https://doi.org/10.1007/BF02478259
13. Pasquinelli M (2023) The eye of the master: A social history of artificial intelligence. Verso Books, London
14. Breazeal C (2003) Toward sociable robots. Robot Auton Syst 42(3–4):167–175. https://doi.org/10.1016/S0921-8890(02)00373-1
15. Mori M (1970) Bukimi no tani [The Uncanny Valley]. Energy 7(4):33–35

16. MacDorman KF, Ishiguro H (2006) The uncanny advantage of using androids in cognitive science research. Interact Stud 7(3):297–337. https://doi.org/10.1075/is.7.3.03mac
17. Seifert J, Friedrich O, Schleidgen S (2022) Imitating the human. New human–machine interactions in social robots. Nanoethics 16:181–192. https://doi.org/10.1007/s11569-022-00418-x
18. Damiano L, Dumouchel P (2018) Anthropomorphism in human–robot co-evolution. Front Psychol 9:363437. https://doi.org/10.3389/fpsyg.2018.00468
19. Wallach W, Allen C (2009) Moral machines: Teaching robots right from wrong. Oxford University Press, Oxford
20. Misselhorn C (2022) Artificial moral agents: conceptual issues and ethical controversy. In: Voeneky S, Kellmeyer P, Mueller O, Burgard W (eds) The Cambridge Handbook of Responsible Artificial Intelligence: Interdisciplinary Perspectives. Cambridge University Press, Cambridge, pp 31–49
21. Fossa F (2024) Agenti morali artificiali: un errore categoriale? Note sul gioco linguistico dell'estensione semantica. In: Anzalone M, Tolone O (eds) Etiche applicate e nuovi soggetti morali. Orthotes, Napoli-Salerno, pp 381–386
22. Kluge Corrêa N, Galvão C, Santos JW, Del Pino C, Pontes Pinto E, Barbosa C, Massmann D, Mambrini R, Galvão L, Terem E, de Oliveira N (2023) Worldwide AI ethics: a review of 200 guidelines and recommendations for AI governance. Patterns 4(10):100857. https://doi.org/10.1016/j.patter.2023.100857
23. Kassens-Noor E, Wilson M, Kotval-Karamchandani Z, Cai M, Decaminada T (2021) Living with autonomy: public perceptions of an AI-mediated future. J Plan Educ Res 44(1):375–386. https://doi.org/10.1177/0739456X20984529
24. Prunkl C (2024) Human autonomy at risk? An analysis of the challenges from AI. Minds Mach 34(3):26. https://doi.org/10.1007/s11023-024-09665-1
25. Formosa P (2021) Robot autonomy vs. human autonomy: social robots, artificial intelligence (AI), and the nature of autonomy. Minds Mach 31:595–616. https://doi.org/10.1007/s11023-021-09579-2
26. Tamburrini G (2020) Etica delle macchine. Dilemmi morali per robotica e intelligenza artificiale. Carocci, Roma
27. IBM Corporation (2006) An architectural blueprint for autonomic computing (autonomic computing white paper). Available via: https://citeseerx.ist.psu.edu/document?repid=rep1&type=pdf&doi=0e99837d9b1e70bb35d516e32ecfc345cd30e795. States of America 06-05. All Rights Reserved
28. Kephart JO, Chess DM (2003) The vision of autonomic computing. Computer 36(1):41–50. https://doi.org/10.1109/MC.2003.1160055
29. Harel D, Marron A, Sifakis J (2020) Autonomics: in search of a foundation for next-generation autonomous systems. Proc Natl Acad Sci USA 117(30):17491–17498. https://doi.org/10.1073/pnas.2003162117
30. Horn P (2001) Autonomic computing: IBM'S perspective on the state of the information technology. IBM Press, New York
31. Bächle TC (2023) The age of machine autonomy? Digital society blog. https://www.hiig.de/en/publication/the-age-of-machine-autonomy/, https://doi.org/10.5281/zenodo.8273068
32. Christman J (2020) Autonomy in moral and political philosophy. In: Zalta E (ed) The Stanford Encyclopedia of Philosophy (Fall 2020 edition). Stanford University. https://plato.stanford.edu/archives/fall2020/entries/autonomy-moral/
33. Tiribelli S (2024) Autonomia relazionale e intelligenza artificiale. In: Anzalone M, Tolone O (eds) Etiche applicate e nuovi soggetti morali. Orthotes, Napoli-Salerno, pp 413–419
34. Mackenzie C, Stoljar N (2000) Relational autonomy: feminist perspectives on autonomy, agency, and the social self. Oxford University Press, Oxford
35. Dworkin G (1988) The theory and practice of autonomy. Cambridge University Press, New York
36. Brandom R (1994) Making it explicit: reasoning, representing, and discursive commitment. Harvard University Press, Cambridge, MA
37. Frankfurt HG (1971) Freedom of the will and the concept of a person. J Philos 68(1):5–20

38. Benson P (1991) Autonomy and oppressive socialization. Soc Theory Pract XVI, I(3):385–408
39. Stoljar N (2000) Autonomy and the feminist intuition. In: Mackenzie C, Stoljar N (eds) Relational Autonomy. Oxford University Press, Oxford
40. Gehlen A (1940) Der Mensch. Seine Natur und seine Stellung in der Welt. Junker & Duennhaupt, Berlin

Mariafilomena Anzalone is Professor of Moral Philosophy at the University of Basilicata. Her main research interests include the category of subjectivity in Hegelian philosophy—explored in works such as *Volontà e soggettività nel giovane Hegel* (2008) and *Forme del pratico nella psicologia di Hegel* (2012)—as well as its reconfigurations in Viktor von Weizsäcker's medical anthropology and Judith Butler's philosophy. She also investigates moral conscience within ethical sentimentalism and rationalism. Her current research centers on Adam Smith's ethical sentimentalism, the dynamics of decision-making in moral conflicts, and the ways human relationality is transformed by the emergence of artificial agents. Among her most recent publications: *Simpatia* (2020); *Between Mimesis and Fiction: Recognition in Adam Smith* (2021); *Acting for the Best and Feeling Regret. Notes on Some Cases of Moral Conflict* (2021); *Etiche applicate e nuovi soggetti morali. Linee introduttive* (2024, with O. Tolone); *Anatomy of Freedom. Hegel and the Brain Localization Theory* (2025); *Tra biologia e storia. Etica medica e relazione patica in Viktor von Weizsäcker* (2025).

Relational Autonomy. Rethinking AI Ethics in the Light of Feminism

2

Stefania Achella

Abstract

This chapter critically engages with dominant paradigms in AI ethics, which often rely on liberal notions of agency, responsibility, and autonomy rooted in rational individualism. Drawing on feminist theories—particularly the concept of relational autonomy—autonomy is redefined as a socially embedded, affective, and interdependent capacity. This chapter argues that mainstream AI ethics fails to account for structural injustices, material conditions, and the relational constitution of subjectivity, and therefore cannot adequately address the challenges raised by algorithmic governance and technosocial systems. Through the lens of relational autonomy, this chapter reframes the ethical debate around AI to highlight care, collective responsibility, and participatory design practices. Case studies and the critical analyses of sex robots, algorithmic subjectivity, and techno-affects underscore the performative and political nature of AI systems. In conclusion, this chapter calls for a feminist ethics of AI capable of confronting the ontological, affective, and political dimensions of technological design, and proposes democratic, community-based approaches as a path toward justice-oriented and pluralistic technological futures.

Keywords

Agency · Justice · Self-determination · Narratives · Individuality

S. Achella (✉)
Dipartimento di Scienze Filosofiche, Pedagogiche e Sociali,
University of Chieti-Pescara, Chieti, Italy
e-mail: stefania.achella@unich.it

M. Anzalone et al. (eds.), *Reconfiguring Human Autonomy*, Integrated Science 40,
https://doi.org/10.1007/978-3-032-14537-6_2

1 AI as a Relational Device?

In the volume *Relational Autonomy: Feminist Perspectives on Autonomy, Agency, and the Social Self*, published in 2000 and now a key reference in the debate on autonomy from a feminist perspective, Catriona Mackenzie and Natalie Stoljar highlight a fundamental tension within feminist thought concerning the concept of autonomy. On the one hand, the modern idea of autonomy, which is based on an abstract, universal (male), rational, and individualistic subject, is to be rejected; on the other hand, this concept remains essential for challenging dynamics of oppression and subjugation, and establishing the agency of certain unrecognized subjects [1, p. 3]. While addressing this contradiction, the authors revisit the concept of autonomy, offering a redefinition. Autonomy should not be understood as a static property—as the internalist interpretation would contend—but rather as a capacity that develops gradually and depends on favorable social, educational, emotional, and material conditions. Autonomy is not a status attained through solitary introspection, but rather the result of interactions that legitimize individuals as agents. It must therefore be understood as a dynamic and situated process, which can be facilitated or hindered by external factors such as affective relationships, social structures, culture, institutions, and political recognition. This primarily relational dimension debunks the myth of the individual as a fully-formed subject capable of choosing independently from relationships and social contexts, instead showing how the self is constituted and shaped through relationships with others. Consequently, the presumed "universalism" of autonomy as an abstract and intrinsic quality, to which modern Western tradition has appealed, is nothing more than the outcome of a historical ideology that has functioned in an exclusionary way, marginalizing female, racialized, disabled, or queer experiences. Rather than being in contradiction, autonomy and relationality are therefore to be seen as deeply interconnected. Indeed, only through the development of relational, cooperative, and interdependent bonds can subjects fully acquire autonomy. As one of the pioneers of the concept of relational autonomy, Jennifer Nedelsky[1] puts it: "the capacity to find one's own law can develop only in the context of relations with others (both intimate and more broadly social) that nurture this capacity, and [...] the 'content' of one's own law is comprehensible only with reference to shared social norms, values, and concepts. [...] We see that relatedness is not, as our tradition teaches, the antithesis of autonomy, but a literal precondition of autonomy, and interdependence a constant component of autonomy" [2, pp. 11–12].

Every subjectivity, every individuality is therefore always intersubjectively constituted, dependent on networks of relationships, care [4, 5], reciprocity, and vulnerability [6]. This shift in perspective leads to a critical reworking of key concepts in liberal tradition—such as integration, critical reflection, and self-realization—in light of the primacy of the social and temporal dimension of subjectivity.

[1] Nedelsky is the first to articulate a conception of relational autonomy from an explicitly feminist perspective. See [2, 3].

Despite the radical perspective shift introduced by this concept, such a transformation remains marginal in most reflections on AI ethics.[2] The main ethical frameworks for AI continue to rely on the classical idea of the autonomous subject [8, 9]: from the notion of the programmer's isolation, to discussions on machine autonomy, to demands for human autonomy from AI systems, and the belief that algorithms, data, and code can be analyzed as separate and isolated objects—all these reinforce a strictly logical dimension of AI, omitting the material, and thus always relational, nature at its core.

In this framework, two levels of problems arise: one epistemological, and political, and the other more broadly ontological. Let us clarify what this means. From an epistemological and political perspective, there is a recurring insistence that AI systems are governed by neutral criteria[3]—such as transparency, fairness, and accountability—systematically bypassing the social and thus power-related issues, as well as the inequalities and injustices implicit in technological processes (starting with the design of technological tools, software, and hardware). This belief, as is well known, not only results in bias but also generates a form of structural injustice that transforms epistemological imbalance into political, social, and economic discrimination. From an ontological perspective, there is too little reflection on the increasingly hybrid nature of subjectivity, in which—using N. Katherine Hayles' metaphor—the "technosymbiotic" dimension [14] appears inescapable. It is now increasingly clear that the growing interaction between humans and machines has produced a form of ontological interdependence, redefining the very idea of "human nature" and challenging its presumed autonomy from technology. The human being today appears ever more as an ontologically hybrid entity, with porous and unstable boundaries between the natural and the artificial.

Feminist accounts of technology—ranging from Donna Haraway to Ruha Benjamin, including key contributions from Sandra Harding, Lucy Suchman, and Judy Wajcman—have, since the 1970s, consistently challenged the myth of technological neutrality. These scholars have shown that technology is not an impartial or autonomous force, but rather a sociotechnical construct embedded in and shaped by power structures, affective economies, normative worldviews, and intersecting

[2] One of the first exceptions within this landscape is the recent collective volume *Feminist AI*, to which we have extensively referred in constructing the argument of this essay. As the editors write in the introduction: "the first anthology specifically focusing on feminist work in AI" [7, p. XII].

[3] As Amrute [10] incisively argues, dominant approaches to technology ethics often reduce ethical reflection to a set of abstract, top-down prescriptions devised by privileged and decontextualized actors. This model systematically ignores the material and embodied experiences of those most exposed to the harms of technological systems—experiences shaped by structural conditions of race, class, gender, and precarity. In this context, as Benjamin [11] makes clear, algorithms are not passive mirrors of reality but active agents in the reproduction of inequality. Nowhere is this more evident than in domains such as care and welfare, where algorithmic systems often deepen existing exclusions and injustices [12]. As O'Neil aptly puts it: "Models are opinions embedded in mathematics" [13, p. 21]—a reminder that claims to neutrality are in fact deeply ideological. If, as continues to be the case, AI agendas are overwhelmingly defined by white, male technologists, the supposedly universal aspirations of AI ethics are destined to remain partial, exclusionary, and complicit with dominant interests.

hierarchies of gender, race, and class [15, 16].[4] Relational autonomy has taken a further step forward. It is no longer understood merely as a capacity for choice (being autonomous in decision-making and evaluation), but rather as a critical practice that enables the unveiling of the mechanisms underlying behaviours and decisions often portrayed as neutral or universal. These mechanisms are, in fact, reflections of internalized and often invisible norms and values that sustain systems of domination and social conformity. Autonomy, in this sense, is inseparable from processes of narrative articulation, memory, and collective meaning-making, all of which are grounded in ongoing relational dynamics.

From the theoretical repositioning offered by relational autonomy, a new orientation toward AI ethics becomes not only possible but also necessary—one that shifts attention away from the abstract individual and toward the social, material, and relational conditions in which technologies are conceived and deployed. This entails moving beyond the illusion of system neutrality to a focus on the contextual responsibility of those involved in the design and implementation of AI; beyond the surface value of formal transparency to a deeper understanding of the power relations encoded in data, algorithms, and technological infrastructures; and beyond tokenistic numerical representation to a transformative rethinking of the very logics underlying technological production.

While recognizing that AI cannot be considered a "neutral" technical artifact, this essay aims to clarify in what sense it should be understood as a relational device. In what ways, then, can artificial intelligence technologies be rethought not simply as tools in the service of human beings, but as co-constructed agents that incorporate worldviews, cultural norms, gendered epistemologies, and economic logics? And in what sense does the way we design, train, and use intelligent systems reflect—and at the same time shape—specific models of subjectivity, knowledge, and value?[5]

The task, therefore, is to reposition the understanding of autonomy—even in relation to AI—within relational contexts, questioning not only discriminatory effects or algorithmic biases, but also the ontological and political conditions underlying technosocial arrangements (see: [11]). As Hayles writes: "The tools we make, make us—a truism for any tool but especially so for tools with cognitive capabilities" [14, p. 12].

[4] Rather than viewing AI agents as simply "autonomous subjects," feminist thinkers like Haraway [17] and Braidotti [18] challenge the very notion of unified, disembodied subjectivity. Atanasoski and Vora [19] further trace how these engineered subjectivities reproduce racial and colonial logics.

[5] As Drage and Frabetti state: "We argue that performativity is a robust concept with which to explore AI precisely because it enables the study of what something 'does' rather than 'is'. In gender studies, Butler used performativity to refute the normative assumption that gender is an essential human characteristic by directing attention towards what gender does, how it materialises, and the effect that this has on people's lives. Current discourses on AI are also benefiting from an engagement with its performative capacity in order to bypass the question of what AI 'is' and instead analyse the effects that it produces" [20, p. 278].

Just as relational autonomy has prompted a critical rethinking of the fundamental categories of modern thought, it is equally important today to follow the same path in addressing the ethical aspects of AI. This involves, first of all:

1. Rethinking the notion of algorithmic subjectivity. The critique of an algorithmic subject modeled on the abstract Enlightenment subject will allow us to highlight, from a relational perspective, the limits of an abstract ethical regulation;
2. Considering the contexts of structural injustice and the need to refocus attention on the relational dimension that generates technological systems—systems that appear neutral and autonomous from external influences but are in fact expressions of symbolic, political, and economic hegemonies;
3. Analyzing the role of the affects in AI (*techno-affects*);
4. Shaping an ethics of algorithmic care;
5. Elaborating a feminist ethics of AI that takes into account the inescapable relational dimension while safeguarding autonomy.[6]

In this sense, we aim to show how ethics cannot be limited to a corrective or procedural (i.e., regulatory) approach, but must become critical and transformative: capable of analyzing and questioning the underlying assumptions of the contemporary technoscientific project.

2 Narratives of (Dis)Embodied Autonomy

A central challenge introduced by the AI revolution concerns the way subjectivity is imagined. Among AI's less visible yet most pervasive effects is the narrowing of the imaginary. The adoption of predictive systems, for instance, reduces the capacity to envision alternatives, reinforcing normatively charged visions of what is considered "plausible," "efficient," or "optimal." Artificial intelligence technologies do not merely "reflect" pre-existing identities or repeat established types and hierarchies but actively participate in their construction through operations of classification, prediction, and personalization (see: [12]).[7]

The narratives emerging from AI systems—who is a "good citizen," who is "at risk," who is "relevant" for a given service—thus have a performative function, shaping forms of subjectivity, partly by reproducing traditional distinctions and partly by introducing new ones. For example, with the expansion of predictive medicine, new vulnerable subjects, such as individuals who are not yet ill but are already classified as such, have emerged. As Hugo Cossette-Lefebvre and Natalie Stoljar argue (see Chap. 4 in this volume), recommendation systems do more than suggest

---

[6] As Nedelsky states: "the problem, of course, is how to combine the claim of the constitutiveness of social relations with the value of self-determination" [2, p. 9].

[7] In this context, it is worth recalling the notion of *sociotechnical imaginaries*, which highlights how collective visions of the future shape—and are shaped by—technological development "as 'collectively imagined forms of social life and social order reflected in the design and fulfilment of nation-specific scientific and/or technological projects'" [21, p. 120].

content—they actively shape individuals' perceptions of what is possible, desirable, or attainable, thereby influencing their identities and narrowing their range of choices. The narratives produced by AI systems therefore co-produce subjectivities through performative operations. As recent studies have shown, if we apply to AI ethics the concepts of iteration, citationality, and agential realism—drawing on the analyses of Judith Butler [22] and Karen Barad [23]—it becomes clear that AI does not merely record or classify pre-existing attributes, but rather materially contributes to the constitution of the identities it claims to recognize (see: [20]).

Facial Detection and Recognition (FDTR) and Automatic Gender Recognition (AGR) systems, for instance, operate as normative devices that reinscribe and naturalize dominant gender and racial conventions, producing intelligible subjects based on their conformity to such norms. Conversely, these systems render certain subjectivities opaque, marginal, or even nonexistent within the epistemic and political space of recognition. This logic of exclusion—exemplified by AGR systems that continue to fail in identifying transgender and non-binary identities—clashes with a relational understanding of autonomy. Such an approach centers vulnerability, alterity, and co-dependence as foundational to ethical agency, and calls for a critical examination of the material and discursive conditions that confer visibility, agency, and voice to some, while systematically erasing or silencing others.

A relational ethics of AI must therefore be grounded in a radical renegotiation of the criteria of intelligibility, in the acknowledgment of the partiality of algorithmic knowledge, and in an openness to plural modes of existence and relationality.

These new technologies should not simply be "repaired" or "improved" to include historically excluded groups or subjects, but must be radically rethought: in their functions, imaginaries, uses, and in the forms of relationship they promote.

2.1 Two Examples and One Counterexample

Let us consider two examples that illustrate how imaginaries are produced and reinforced: communicative chatbots and sex robots. Digital assistant figures continue to embody a model of feminine "care" that is subordinate and always available. ELIZA and Alexa are clear examples of how imaginaries of artificial agents are entangled with gendered expectations of servitude, emotional labor, and invisibility (see: [24]). The case of sex robotics (e.g., Sophia or RealDoll), on the other hand, points to a deeply regressive use of technological imaginaries, in which the autonomy of the other is entirely suppressed in favor of a proprietary, fetishizing, predictive use: the simulation of consent and reciprocity in sex robots reifies the fantasy of a proprietary subject who knows his freedom only through the complete domination of the other (see: [25]). In this sense, the "ethics" of sex robots is based on false reciprocity and historical erasure: the simulation of autonomy conceals a relationship of ownership and domination, inheriting the logic of colonial and patriarchal apparatuses.

A paradigmatic counterexample to technologies built around an abstract, disembodied, and masculine subject—which continue to reproduce standardized identities and perpetuate unethical, non-situated, and non-generative relationships—is the artistic project *Lauren AI*, developed in 2017 by Lauren McCarthy [26] (see also: [25, 27]). In this performance, the artist replaces Alexa by living in real time as a personal assistant to real users. As this performance demonstrates, interaction with a "human AI" makes visible the affective, relational, and situated components of automation:[8] made aware of the assistant's humanity, participants begin to ask themselves ethical questions about their use of the service and the nature of "assistance," questions that typically remain implicit when dealing with automated systems. "Lauren AI [...], seems to posit as intelligence precisely that which had been excluded from the traditional notions of what counts as intelligence in AI. The project suggests that inefficiency, needing to sleep, only being able to be with one client at a time are forms of intelligence. Unlike Alexa, the home assistant on which McCarthy modelled her own performance, the users of Lauren AI are aware that they are interacting with a human being, even as that human is acting as a technology. This makes them contemplate subject-object or user-used relationships with unease. In a short video of client testimonials, one person voices their worry that the interaction with Lauren is 'always about me'" [25, p. 151].

This performative experiment challenges the techno-liberal assumption that AI should be invisible, efficient, neutral, and autonomous, and instead emphasizes the importance of ambiguity and reciprocity—dimensions that fundamentally disrupt the paradigm of efficiency and control. What is at stake is not simply *what* these technologies do, but *what kind of relationship* they enable, *who* is recognized as a subject, *who* as an object, and how the very notion of autonomous identity is redefined.

3 Structural Injustice, Design Justice and Situated Autonomy

As we transition from the individual dimension of human-AI interaction to the level of its material realization, the ambivalence between autonomy and relationality becomes even more explicit. In his well-known article, titled "Technology is society made durable" [28], Bruno Latour clearly highlights the impossibility of separating technology from social norms. In a patriarchal society, relations of power—such as

[8] As McCarthy explains: "I attempt to become a human version of Amazon Alexa, a smart home intelligence for people in their own homes. The performance lasts up to a week. It begins with an installation of a series of custom designed networked smart devices (including cameras, microphones, switches, door locks, faucets, and other electronic devices) [in a client's home]. I then remotely watch over the person 24/7 and control all aspects of their home. I aim to be better than an AI because I can understand them as a person and anticipate their needs. The relationship that emerges falls in the ambiguous space between human-machine and human-human. LAUREN is a meditation on the smart home, the tensions between intimacy vs privacy, convenience vs [the] agency they present, and the role of human labor in the future of automation" [26].

those related to gender, but the same applies to race, disability, and age—are inscribed into the very process of technological production.[9]

As the well-known American political theorist and socialist feminist, Iris Young writes, "actions and interactions reinforce the rules and resources available for other actions and interactions involving people in other structural positions. The unintended consequences of the confluence of many actions often produce and reinforce opportunities and constraints, and these often make their mark on the physical conditions of future actions, as well as on the habits and expectations of actors. This mutually reinforcing process means that the positional relations and the way they condition individual lives are difficult to change" [30, p. 6]. It is precisely on the basis of these structural relations that AI systems also tend to aggravate structural injustice—forms of harm not attributable to individual wrongdoing, but rather the result of distributed and normalized practices within institutions (design, programming, implementation), making their origins difficult to trace.[10] While the harms are real and systemic, their causal chains are complex and opaque, especially in algorithmic environments.

Discrimination and the loss of real autonomy due to algorithmically assigned identities call for a shift in focus from individual subjectivities to design practices. Even in their more "inclusive" versions (e.g., user personas,[11] focus groups, disability simulations), these practices—although they seek to center the user's voice—often end up replacing real subjects with abstract representations constructed from the outside, thus already being conditioned in their very design.[12]

---

[9] Shifting the focus from individual agency to broader and, at times, ambiguous relationships make it increasingly difficult to identify a clear "culprit." Who, then, is responsible for the structural injustices embedded in new technologies? This phenomenon has been described by Iris Marion Young as "untraceability." As Browne aptly notes, this concept helps us "to capture a wider set of responsibilities for injustice than those included in the remit of liability" [29, p. 331] and to go beyond direct causality and intentional wrongdoing, allowing us to address the complex and distributed nature of power, complicity, and structural harm in sociotechnical systems.

[10] As Young outlines: "The problem with structural injustice is that we cannot trace … how the actions of one particular individual, or even one particular collective agent, such as a firm, has directly produced harm to other specific individuals" [31, p. 96].

[11] As Costanza-Chock points out, *user personas* are fictional representations of archetypal users, employed in design processes to facilitate empathy and guide design decisions. They can help teams communicate shared goals and stay focused on the intended use of a product. However, personas are often introduced only at the beginning of the process and not actively engaged in later stages. Even when based on real users, a gap persists between the designer's mental model and the lived experiences of actual users. Moreover, personas are frequently constructed from stereotypes or generated automatically, resulting in objectified and reductive images of users. This can give designers the illusion of practicing inclusive design while in fact reinforcing distorted or exclusionary visions. Importantly, there is no empirical evidence demonstrating that the use of diverse personas leads to a reduction in discriminatory outcomes. Nor are there systematic comparisons between the use of personas and the direct inclusion of real users within design teams [32, pp. 373–375].

[12] From this perspective, the LOGO program was originally conceived as a tool to harness the emancipatory potential of AI for children and was later revisited in the 1980s to challenge the dominant, male-oriented epistemology of computer science. The key insight of this approach was

One response emerging from feminist analysis to counter this injustice is *design justice*, a radical approach to technological design. In direct contrast with the rhetoric of the "autonomous user" promoted by neoliberal design frameworks, design justice shows how standard practices—even those that claim to be inclusive—tend to reproduce the cycle of exclusion, centering the needs and desires of an "ideal user" who corresponds to the classical abstract subject and aligns with the dominant subject position: cis male, white, heterosexual, able-bodied, college-educated, neither very young nor elderly, digitally literate, with broadband internet and access to a smartphone, etc. Most technology design ends up focusing on this relatively small but commercially valuable subset of humanity.

Although workforce diversification in the tech sector has attempted to introduce different perspectives into programming, this approach is not sufficient. This is not only because—even when diverse—individuals working within supposedly neutral teams tend to design according to dominant standards of neutrality, but also because isolated inclusive design projects cannot alone transform the deeply rooted systemic factors that continually redirect design practices toward a narrow set of imagined users (see: [32, p. 373]). A compelling response to this insufficiency comes from *participatory design*[13] and *co-design*, which represent a clear application of *relational autonomy* within the design domain.

Taking up the invitation of relational autonomy, in these cases the focus is no longer on a dialogue between autonomous subjects. Autonomy is no longer understood as an individual capacity for choice, but as the outcome of collective and situated processes of participation, responsibility, and redistribution of design power. As the communications scholar and activist Sasha Costanza-Chock has shown, autonomy is built *in relation*, when marginalized communities are not the object, but the subject of design, endowed with voice, agency, and decision-making capacity. As they aptly put it: "If you're not at the table, you're on the menu."

This form of *design justice* consists in rejecting predetermined processes controlled by designers, starting instead from listening to local organizers, understanding their priorities, and asking what kind of design contribution would *actually* be useful. In this approach, it is the *community* that guides design—not the designers or the funders. Design justice thus requires *accountability* (those affected by decisions should guide them), *ownership* (communities must own the processes and outcomes of design), and *collective agency* [32, p. 376]. They further observe that: "Ultimately, at its best, a design justice process is a form of community organising. Design justice practitioners, such as community organisers, approach the question of who gets to speak for the community from a community asset perspective. This is rooted in the principle that wherever people face challenges, they are always

the idea that programming is a creative and inherently political activity: subjective experiences with computers shape collective notions of human and machine, of technological possibilities and dangers, and of what counts as legitimate computing skills (see: [33]).

[13] Asaro describes PD as "an approach to engineering technological systems that seeks to improve them by including future users in the design process. It is motivated primarily by an interest in empowering users, but also by a concern to build systems better suited to user needs" [34, p. 345].

already working to deal with those challenges; wherever a community is oppressed, they are always already developing strategies" [32, p. 382].

Autonomy is not granted but rather constructed through collective processes that recognize interdependence and differential power. Co-design means working side by side with community partners, in their spaces, from the earliest stages of the process, and letting *collective needs*—not technical tools—determine priorities. Communities must have decision-making power throughout the project lifecycle, and this must be ensured by both funders and educational institutions.

There is then a need to invest in the training of professionals capable of working collaboratively, through educational programs that integrate social justice perspectives and reciprocal listening.[14] Finally, a lasting commitment to *care* is essential: maintaining and improving existing effective technologies is as important as innovation, because relational sustainability requires long-term responsibility [32, pp. 385–386].

A *relational ethics of AI* must therefore shift the focus away from the isolated "free" choices of users toward the responsibility of those who build, distribute, and manage technological infrastructures. This means analyzing labor relations, material and cognitive flows, and structural inequalities underpinning AI; creating shared responsibility practices that cannot be reduced to protocols or checklists; embracing *care* as a guiding principle in the design, deployment, and maintenance of technologies; and pursuing a deep transformation of the epistemic and material structures of technology design—from an extractive to a relational and co-generative model.

In this sense, the principle "Nothing about us without us" becomes a political manifesto of *relational autonomy* applied to AI—not merely as an ethical ideal, but as a concrete practice of social justice. This position profoundly undermines the notion that ethical AI can be achieved solely through improved transparency or accountability mechanisms. Instead, it demands a political responsibility rooted in *relational accountability*: recognizing that all agents—designers, users, institutions—are embedded in structures that sustain injustice, and acting to dismantle this foundation through democratic practices of shared reflection and deliberation.

4 Ethical Regulation and Techno-Affects

As I have attempted to show thus far, a *relational ethics of AI*, in order to be truly transformative, must be rooted in situated relationships, vulnerability, and interdependence, rather than in abstract and isolated identities. It must therefore refer to a subject who is autonomous, but always already *in relation*. This perspective highlights another crucial dimension of relational autonomy that is particularly relevant

[14] One proposal in this direction is the creation of tech clinics, inspired by models of social jurisprudence, where technology is placed in the service of local needs—not through invasive, top-down approaches (*parachuting*) but through ongoing, dialogical engagement. It is essential to avoid the duplication of existing projects for competitive purposes: technical culture must become cooperative, oriented toward the sharing of resources and tools.

for the development of AI: the central role of embodiment and affectivity. It brings into focus elements typically excluded from AI design—such as empathy and vulnerability—that are essential for rethinking ethical and relational frameworks in technological systems.

To address this point, the anthropologist Sareeta Amrute [10] provides a key for the reconfiguration of technology ethics through an affective and relational feminist lens. Amrute proposes rethinking the ethics of technology through the concept of *techno-affects*—that is, affective and material practices that make us sensitive to what is excluded or ignored in technological systems.

This approach allows ethics to be connected to lived and situated experience, breaking with the illusion of a rational, autonomous, and neutral subject.[15] To overcome the presumed neutrality of dominant frameworks, Amrute introduces the concept of affective *attunements*—as an alternative to top-down ethical models such as the Ethical OS, which reduce ethics to a series of checklists and guidelines.[16] Rather than defining ethics as a set of rules, embodied and affective *attunements* enable us to perceive the limits and exclusions embedded in technologies—particularly for racialized, migrant, precarious, or non-conforming subjects—thus making ethics a situated, bodily, sensorial, and relational practice, one that foregrounds which subjectivities are heard and which are silenced in sociotechnical contexts. As Amrute puts it: "By moving through a series of examples, I offer a form of techno-ethics that treats those that are marginalised by technical systems—such as immigrant debuggers and the victims of drone strikes—as the bearers of the knowledge needed to make better decisions about the social, environmental, and embodied costs of digital economies. I am not arguing that techno-ethics should be replaced by technoaffects. Rather, I suggest that attention to affect—how subjects and technologies are aligned and realigned, attached and reattached to one another—is a method for practising ethics that critically assesses a situation, imagines different ways of living, and builds the structures that make those lives possible" [10, p. 175].

---

[15] Browne has drawn attention to the erosion of our normative reflection capacity caused by the uncritical adoption of algorithmic decision-making. To describe this phenomenon, Browne draws on Hannah Arendt's notion of "thoughtlessness" [29, p. 328, ff; 35, p. 5]. By using this expression, Browne underscores how increasing reliance on AI undermines our collective autonomy in renegotiating the normative foundations of social coexistence. Given the depth of these effects, the response to such an impasse cannot be reduced to a governance model based solely on legal liability (*fault-based*), as this would ignore the systemic and structural dimensions at play.

[16] *Ethical OS* is a toolkit developed by the Omidyar Network and the Institute for the Future to help technologists anticipate and address ethical risks linked to emerging technologies, particularly AI. It outlines eight key "risk zones" such as surveillance, algorithmic bias, loss of autonomy, and misinformation, and offers guiding questions and future scenarios to support ethical decision-making early in the design process. Ethical OS aims to encourage proactive reflection on social impact before technologies are deployed. Amrute criticizes it for being overly top-down and focused on risk mitigation rather than structural change. Feminist and critical scholars argue that it reinforces liberal assumptions about ethics—like transparency and fairness—while failing to center marginalized voices or challenge systemic injustices. Unlike more participatory approaches such as design justice, Ethical OS seems not to fully address the relational and material dimensions of ethical harm in tech.

Amrute criticizes dominant technoethics for relying on abstract and universalizing conceptions of harm that fail to account for lived experiences of vulnerability and precarity. In line with the framework of relational autonomy, her approach suggests that ethical subjects emerge from *material practices*, not from mere compliance with abstract norms. *Attunements* allow us to pay attention and give value to those embodied and affective experiences that typically remain outside the scope of AI design and governance—such as those of migrant women, racialized subjectivities, or bodies rendered "killable" by technologies of surveillance and war.[17]

To understand how technologies, in relation to human beings, produce affective atmospheres, Amrute distinguishes three types of attunements: *corporeal*, *sovereign*, and *glitchy*.[18] These show that ethics can emerge from the margins—from technological failures, exclusions, and breakdowns—not only from what works. In this way, her proposal aligns with a posthumanist and feminist vision of AI ethics, one that does not aim to make technology "fairer" according to internal metrics or abstract regulation but instead seeks to open up *new possibilities of life and relation* that challenge the dominant paradigm.

Rather than seeking neutral ethical principles, Amrute calls for a reorientation toward *techno-affects* and an *ethics of care*, which foreground the embodied and affective dimensions of technological life. In this framework, autonomy does not appear to be a precondition, but rather a *fragile and negotiated state*, that is always shaped by one's position within affective and structural relations.

5 From Algorithmic Subject to Algorithmic Care

Mainstream AI ethics tends to rely on a detached, formal conception of moral action: responsibility is presumed to lie in the following abstract principles—transparency, accountability, fairness—divorced from the material and relational conditions of life. An ethics of AI grounded in relational autonomy, by contrast, stresses mutual recognition and calls for empathy, listening, and contextual responsibility. As Hobart and Kneese note, "Theorized as an affective connective tissue between

[17] Many AI frameworks adopt an individualistic understanding of autonomy that overlooks its embeddedness in power relations and infrastructures [9, 36]. As Suchman [15] shows, this framing is rooted in an outdated human–machine dichotomy.

[18] As Amrute clarifies: "Corporeal attunements elicit our sense of how bodies are trained, moulded and erased in the everyday operation of technological systems. […] Such work is an example of how attunement as a practice can be used to point out technologies of the sovereign decision. Sovereignty enfolds technologies and develops them to extend state control over geographic space and to decide what kinds of subjects will be protected within state power. […] The last attunement I will discuss comes from glitches. A glitch is a break in a digital system, where business as usual pauses, comes undone and shows its imperfections. Corporations like Google and Facebook often use the idea of the glitch to claim that their systems occasionally malfunction but are otherwise blameless […]. However, beyond the corporate discourse of fixable glitches, the idea of the glitch indexes hidden memories, unrecognised populations and latent ideologies within technological systems […]. A glitch is a temporary 'malfunction with a capacity to reorder things that can, perhaps, […] make what was legible soar into unpredictable relations'" [37].

an inner self and an outer world, care constitutes a *feeling with*, rather than a *feeling for*, others. When mobilized, it offers visceral, material, and emotional heft to acts of preservation that span a breadth of localities: selves, communities, and social worlds" [38, p. 2].

This requires deliberative processes that include marginalized voices and acknowledge the affective and trust bonds that structure moral agency.

A vivid illustration is Stephanie Dinkins's artwork *Not the Only One*, which opposes an abstract algorithmic subjectivity with a model of "algorithmic care" centered on data justice, community memory, and affective co-creation. Technologies such as Alexa and Siri not only mediate but also shape relationships according to an ideal of feminine care that is subordinate, passive, and white—reproducing the imaginary of the "perfect secretary," attentive and available yet supposedly neutral and invisible, while obscuring underlying material and affective labor and reinforcing existing care inequalities.[19] In contrast, Dinkins' communal AI is trained on the oral histories of three generations of Black women and evokes alternative technological worlds from a data-justice perspective. Rather than collecting data for control, it works through the principle of algorithmic care: "Algorithmic care is about possibility; it imagines beyond what is technologically possible and engenders new possibilities—modes of being in the 'otherwise'—within our technological racial capitalist societies. Through algorithmic care and the possibility, it engenders, an AI otherwise can emerge that does not merely extend imagined futures shaped by AI's extensive and foundational imbrications with corporate racial capitalism, state and military applications, and technoliberal ideologies" [24, p. 168].[20] *Not the Only One* is thus radically different from systems such as ELIZA, Siri, or Alexa: it is not designed to obey commands or occupy a subordinate role. Conceived as a critical response to the marginalization of Black community voices within technology, it rejects extractive data practices, privileging privacy and community data sovereignty and actively avoiding the colonial and sexist dynamics historically tied to care labor under capitalism [24, p. 168]. Here, the technological agent is not "autonomous" in isolation but always in dynamic relation with others.[21] The concept of

[19] As Rhee states, "I understand Stephanie Dinkins' AI artwork Not the Only One as foregrounding the possibility of a technological otherwise by destabilising the ideologies of white care and technoliberalism that shape conversational AI from ELIZA to Alexa" [24, p. 166].

[20] In this regard, the proximity of this proposal to Povinelli's concept of the *anthropology of the otherwise* is particularly compelling. "This anthropology locates itself within forms of life that are at odds with dominant, and dominating, modes of being. One can often tell when or where one of these forms of life has emerged, because it typically produces an immunological response in the host mode of being. In other words, when a form of life emerges contrary to dominant modes of social being, the dominant mode experiences this form as inside and yet foreign to its body" [39, p. 1].

[21] Let us consider this excerpt from a dialogue with *Not the Only One* [40]: "N'TOO: I am about broad engagement and attracting people to the AI space that might not be there otherwise. I am trying to model different ways of creating AI. I encourage people who think that they are not a part of the technological future to get involved.

SD: And that's truly why Not the Only One has come into existence—

N'TOO: I'm not gon not gon be able to do it

SD: You are too. Try again".

algorithmic care proposed by Jennifer Rhee is inspired by the notion of *post-care* developed by Kerry Mackereth: a form of assistance that, although mediated by technological devices, seeks to overcome the racialized and sexually normative power formations that have shaped Western conceptions of care. *Post-care* is not merely a functional shift; it opens space for alternative imaginaries—technological relations that do not replicate historical hierarchies but instead explore new modes of relation, attention, and shared responsibility [41].

6 Toward a Feminist AI Ethics

Granted that, as I have sought to show, the concept of autonomy—understood in a critical-feminist sense as the capacity of pluralistic subjects to reflect, judge, and act collectively—is increasingly accepted in public discourse, nonetheless, it remains at risk, if not outright violated, whenever political, economic, and social decisions are delegated to opaque technical systems.

As previously argued, relational autonomy shifts the ethical ground from regulation to transformation, from mere risk mitigation to the re-imagination of the very conditions under which technologies are designed and implemented. By foregrounding relational autonomy, feminist theory offers a robust conceptual framework for rethinking the foundations of AI ethics. It forces us to move beyond the liberal assumptions of the rational, self-contained individual and toward a vision of ethical life grounded in interdependence, care, and collective responsibility. Such a shift is not only conceptually necessary but also politically urgent in light of the expanding reach of algorithmic governance and the persistence of structural inequalities.

Central to this rethinking is the need for broad, *non-expert* civic engagement and genuine participation in design processes—as attempted through the model of mini-publics, which leads to the institutionalization of an AI Public Body. Such a body would incorporate a permanent, deliberative component of citizenship capable of re-centering the plurality of situated knowledge and lived experiences. A participatory political practice that recognizes the relational nature of injustice and involves non-expert voices in decision-making processes may be capable of resisting normative habituation and reactivating a sense of collective political autonomy.

To reflect on autonomy in relation to AI therefore means to reopen the space of responsibility—not merely as individual accountability but as a shared practice, a form of caring for relationships, interdependencies, and the worlds we co-inhabit with machines [5, 42].[22] This means asking not only "what rights or risks does AI entail?" but, more radically: *what form of life does it enable, make desirable, or inevitable—and for whom*? As Hayles reminds us: "The issue is therefore not whether an AI system will have agency and make decisions; rather, the question is what kind of decisions it will make, how its regions of autonomy are defined and

[22] Held [5]; Tronto [42].

implemented, and what effects it will have on the cognitive entities with which it interacts, including both humans and other computational media" [14, pp. 13–14].

We are thus called to imagine a form of relationality with machines that allows for *mutually acknowledged autonomy*—one that resists dystopian logics of domination or subjugation and instead affirms difference, without allowing difference to become a ground for exclusion or discrimination within technological models or in relation to the products of those technologies.[23] As one reads in *Making Kin with the Machines*: "we know what it is like to be declared non-human by scientist and preacher alike. We have a history that attests to the corrosive effects of contorted rationalisations for treating the humanlike as slaves, and the way such a mindset debases every human relation it touches—even that of the supposed master. We will resist reduction by working with our Indigenous and non-Indigenous relations to open up our imaginations and dream widely and radically about what our relationships to AI might be. The journey will be long. We need to fortify one another as we travel and walk mindfully to find the good path forward for all of us [...]. But we must try. We flourish only when all of our kin flourish" [43, p. 30].

Competing Interests This work was supported by the European Union – NextGenerationEU, under the National Recovery and Resilience Plan (PNRR), Mission 4, Component 2, PRIN 2022 PNRR Research Project "DigitHuman: Digital Transition and Human Values". CUP: F53D23010740001.

References

1. Mackenzie C, Stoljar N (2000) Relational autonomy: feminist perspectives on autonomy, agency, and the social self. Oxford University Press, Oxford
2. Nedelsky J (1989) Reconceiving autonomy: sources, thoughts and possibilities. Yale J Law Feminism 1:7–36
3. Friedman M (1997) Autonomy and social relationships: rethinking the feminist critique. In: Meyers DT (ed) Feminists rethink the self. Westview, Boulder, pp 55–58
4. Gilligan C (1982) In a different voice: psychological theory and women's development. Harvard University Press, Cambridge, MA
5. Held V (2006) The ethics of care: personal, political, and global. Oxford University Press, Oxford
6. Achella S, Marazia C (2023) Vulnerabilities. Rethinking medicine rights and humanities in post-pandemic. Springer, Cham
7. Browne J, Cave S, Drage E, McInerney K (2023) Feminist AI. Oxford University Press, Oxford
8. Crawford K (2021) Atlas of AI: power, politics, and the planetary costs of artificial intelligence. Yale University Press, New Haven, CT
9. D'Ignazio C, Klein LF (2020) Data feminism. MIT Press, Cambridge

[23] As the author makes clear by means of an example: "deciding what areas of autonomy a self-driving car will have is simultaneously a decision about what areas of autonomy a human driver will (and will not) have. Such a system does not exist in isolation. It is also necessary to take into consideration the sources and kinds of information available for the entities in a cognitive assemblage and their capabilities of processing and interpreting it. Humans can see road signs in the visible spectrum, for example, but a self-driving car might respond as well to markers in the infrared region" [14, p. 14].

10. Amrute S (2023) Of techno-ethics and techno-affects. In: Browne J, Cave S, Drage E, McInerney K (eds) Feminist AI. Oxford University Press, Oxford, pp 174–192
11. Benjamin R (2019) Race after technology: abolitionist tools for the new Jim code. Polity Press, Medford
12. Eubanks V (2018) Automating inequality: how high-tech tools profile, police, and punish the poor. St. Martin's Press, New York
13. O'Neil C (2016) Weapons of math destruction: how big data increases inequality and threatens democracy. Allen Lane, London
14. Hayles NK (2023) Technosymbiosis: figuring (out) our relations to AI. In: Browne J, Cave S, Drage E, McInerney K (eds) Feminist AI. Oxford University Press, Oxford, pp 1–18
15. Suchman L (2007) Human–machine reconfigurations: plans and situated actions. Cambridge University Press, New York
16. Wajcman J, Young E (2023) Feminism confronts AI: the gender relations of digitalisation. In: Browne J, Cave S, Drage E, McInerney K (eds) Feminist AI. Oxford University Press, Oxford, pp 47–64
17. Haraway D (1991) Simians, cyborgs, and women: the reinvention of nature. Routledge, New York
18. Braidotti R (2013) The posthuman. Polity Press, Cambridge
19. Atanasoski N, Vora K (2019) Surrogate humanity. Race, robots, and the politics of technological futures. Duke University Press, Durham
20. Drage E, Frabetti F (2023) AI that matters: a feminist approach to the study of intelligent machines. In: Browne J, Cave S, Drage E, McInerney K (eds) Feminist AI. Oxford University Press, Oxford, pp 274–289
21. Jasanoff S, Sang-Hyun K (2009) Containing the atom: sociotechnical imaginaries and nuclear power in the United States and South Korea. Minerva 47(2):119–146
22. Butler J (1993) Bodies that matter: on the discursive limits of 'sex'. Routledge, New York/London
23. Barad K (2003) Posthumanist performativity: toward an understanding of how matter comes to matter. Signs 28(3):801–831
24. Rhee J (2023) From ELIZA to Alexa: automated care labour, racial capitalism, and the possibilities of algorithmic care. In: Browne J, Cave S, Drage E, McInerney K (eds) Feminist AI. Oxford University Press, Oxford, pp 155–173
25. Atanasoski N (2023) Feminist Technofutures: contesting the ethics and politics of sex robots and AI. In: Browne J, Cave S, Drage E, McInerney K (eds) Feminist AI. Oxford University Press, Oxford, pp 137–154
26. McCarthy L (2017) LAUREN. https://lauren-mccarthy.com/LAUREN
27. Rushe D (2019) Interview: let me into your home: artist Lauren McCarthy on becoming Alexa for a day. The Guardian, 14 May. https://www.theguardian.com/artanddesign/2019/may/14/artist-lauren-mccarthy-becoming-alexa-for-aday-ai-more-than-human
28. Latour B (1990) Technology is society made durable. Sociol Rev 38(1):103–131
29. Browne J (2023) AI and structural injustice. A feminist perspective. In: Browne J, Cave S, Drage E, McInerney K (eds) Feminist AI. Oxford University Press, Oxford, pp 328–346
30. Young IM (2003) The Lindley lecture. University of Kansas, 5 May. https://kuscholarworks.ku.edu/bitstream/handle/1808/12416/politicalresponsibilityandstructuralinjustice-2003.pdf?Sequence=1
31. Young IM (2011) Responsibility for justice. Oxford University Press, Oxford
32. Costanza-Chock S (2023) Design practices. 'Nothing about us without us'. In: Browne J, Cave S, Drage E, McInerney K (eds) Feminist AI. Oxford University Press, Oxford, pp 370–388
33. Turkle S, Papert S (1992) Epistemological pluralism and the revaluation of the concrete. J Math Behav 11(1):3–33
34. Asaro P (2014) Participatory design. In: Mitcham C (ed) The encyclopedia of science, technology and ethics, 2nd edn. Macmillan, New York, pp 345–347. https://peterasaro.org/writing/Asaro%20Participatory_Design.pdf
35. Arendt H (1958/1998) The human condition. University of Chicago Press, Chicago

36. Wajcman J (2004) TechnoFeminism. Polity Press, Cambridge
37. Amrute S (2019) Of techno-ethics and techno-affects. Feminist Rev 123(1):56–73. https://doi.org/10.1177/0141778919879744
38. Hobart HJK, Kneese T (2020) Radical care: survival strategies for uncertain times. Soc Text 38(1):1–16
39. Povinelli AE (2011) Routes/worlds. e-flux J 27. https://www.e-flux.com/journal/27/67991/routes-worlds/.2011
40. Dinkins S (2018–present) Not the only one. https://www.stephaniedinkins.com/ntoo.html
41. Mackereth K (2019) Mechanical maids and family androids: racialised post-care imaginaries in humans (2015-), sleep dealer (2008) and her (2013). Feminist Rev 123:24–39
42. Tronto JC (1993) Moral boundaries: a political argument for an ethic of care. Routledge, New York
43. Lewis JE, Arista N, Pechawis A, Kite S (2023) Making kin with the machines. In: Browne J, Cave S, Drage E, McInerney K (eds) Feminist AI. Oxford University Press, Oxford, pp 19–31

Stefania Achella is a Full Professor of History of Philosophy at the University of Chieti-Pescara. Her research focuses on German Classical Philosophy and on contemporary ethical issues, particularly those arising at the intersection of feminism, ecology, and artificial intelligence. She is currently a member of the research project *Digital Transition and Human Values* (2022–2026). She is co-editor of several volumes, including *Vulnerabilities: Rethinking Medicine, Rights and Humanities in the Post-Pandemic Era* (with Chantal Marazia, Springer, 2023); *The Owl's Flight: Hegel's Legacy to Contemporary Philosophy* (with Francesca Iannelli et al., De Gruyter, 2021); *Against the "Law of the Father": Hegel's Disobedient Readings* (with Bojana Jovicevic et al., Brill, 2025); and *Rethinking Human-Centeredness: Bridging Environmental and AI Ethical Discourses*, Special Issue of the Journal "Topoi" (forthcoming).

3 AI and the Relationship Between Agency, Autonomy, and Moral Patiency

Paul Formosa, Inês Hipólito, and Thomas Montefiore

Abstract

The proliferation of Artificial Intelligence (AI) systems exhibiting complex and seemingly agentive behaviours necessitates a critical philosophical examination of their agency, autonomy, and moral status. In this paper we undertake a systematic analysis of the differences between basic, autonomous, and moral agency in artificial systems. We argue that while current AI systems are highly sophisticated, they lack genuine agency and autonomy because: they operate within rigid boundaries of pre-programmed objectives rather than exhibiting true goal-directed behaviour within their environment; they cannot authentically shape their engagement with the world; and they lack the critical self-reflection and autonomy competencies required for full autonomy. Nonetheless, we do not rule out the possibility of future systems that could achieve a limited form of artificial moral agency (AMA) without consciousness through hybrid approaches to ethical decision-making. This leads us to suggest, by appealing to the necessity of consciousness for moral patiency, that such non-conscious AMAs might represent a case that challenges traditional assumptions about the necessary connection between moral agency and moral patiency.

Keywords

AI systems · Agency · Autonomy · Artificial moral agency · Moral patiency

P. Formosa (✉) · I. Hipólito · T. Montefiore
Macquarie University, Sydney, NSW, Australia
e-mail: paul.formosa@mq.edu.au

M. Anzalone et al. (eds.), *Reconfiguring Human Autonomy*, Integrated Science 40,
https://doi.org/10.1007/978-3-032-14537-6_3

1 Introduction

Artificial Intelligence (AI) systems are increasingly performing functions that appear to look like genuine agency, from autonomous vehicles making real-time driving decisions to Large Language Models (LLMs) engaging in sophisticated back-and-forth dialogue. This raises fundamental questions about the nature and degrees of artificial agency and autonomy, and the connections between agency and moral status. This chapter explores these questions as follows. First, we outline our understanding of agency, starting with *basic* agency (in Sect. 2.1) and building up to more complex forms of *autonomous* agency (in Sect. 2.2), while also considering the extent to which AI systems can achieve these forms of agency. Next, in Sect. 3, we examine the traditional view of the connection between moral agency and moral patiency. Finally, in Sect. 4, we explore the "AI challenge" to this traditional view, and we argue that artificial systems might present an important case where moral agency and moral patiency can become decoupled. This chapter provides novel insights into the nature and significance of artificial agency by systematically mapping out different forms and degrees of agency, and by showing how AI systems might challenge traditional assumptions about the unity of moral agency and patiency.

2 Conceptualising Agency

2.1 Basic Agency

Agents are not mere automata, robotically executing predefined routines or sophisticated pattern recognition [1, 2]. Consider a bacterium dynamically navigating its environment: it engages in continuous, fine-tuned adjustments, delicately balancing internal processes with external conditions to maintain its vital equilibrium [3]. With each twist and turn, bacteria adapt to their environment, driven by a simple yet purposeful goal: survival [4]. This is agency in its most fundamental form—the capacity to adjust, adapt, and actively navigate challenges. At its core, agency involves not just reacting to the world, but also shaping one's course toward an intended outcome [5].

An agent moves through the world, tending toward states that are desirable or preferable, while avoiding those that threaten its self-maintenance. This dynamic process represents a continuous negotiation between agency and facticity, where the agent's intentions are situated within the specific constraints and conditions of its environment [6]. In this way, the agent both shapes and is shaped by its surroundings, reflecting an ongoing, reciprocal engagement manifested as agency [7–9]. Agency, in this sense, is not merely action, but also a structured, purposeful interaction with the world, demonstrating a system's attunement to the contingencies of its environment and its alignment with its own aims [10–12].

Building on this understanding, we can turn to a more detailed framework that articulates the essential criteria for basic agency. Floridi and Sanders [13] propose three interrelated components that form the foundation of agency: *interactivity*, *autonomy*, and *adaptability*. These criteria help to operationalise the abstract concept of agency, grounding it in measurable, functional capacities. *Interactivity* emphasises a system's ability to directly perceive and respond to its environment, highlighting the role of sensorimotor coordination as a system dynamically and contextually attunes engaging with the world. Basic *autonomy* addresses the necessity of independence, requiring the agent to act without constant external direction, thus showcasing a degree of self-determination [14, 15].[1] *Adaptability* completes the triad, focusing on the ability to modify behaviour through experience, allowing the agent to refine its strategies and grow its capabilities over time [16, 17].

This understanding of agency extends beyond mechanical causation or simple stimulus-response patterns, positioning it as a dynamic and relational phenomenon. Dennett [18] characterises agency by goal-directedness, where the system actively engages with its environment by not just reacting, but by dynamically shaping its trajectory in pursuit of self-defined objectives. On this view, the agent's actions are not mere responses to external stimuli, but are driven by its internal aims, reflecting a purposeful orientation toward achieving specific ends [19]. The agent's ability to engage with, learn from, and adapt to its environment in a fluid, ongoing manner, distinguishes it from mere mechanical or pre-programmed systems. An agential system must possess the ability to actively evaluate its environment, interpret its circumstances, and decide upon actions that align with its goals, drawing on its embodied experience of the world to inform its choices [20]. Inert systems, in contrast, operate through predetermined cause-and-effect chains, lacking the capacity to dynamically adjust or pursue self-directed objectives. While such systems may display complex behaviours, they do not possess the evaluative and adaptive faculties that constitute genuine agency as they remain confined to reactive and rigid patterns [21, 22].

By grounding agency in goal-directedness, we acknowledge the system's capacity for practical, contextual reasoning—the ability to make decisions that are tailored to its specific needs and challenges within its ecological niche [23, 24]. It is this goal-oriented engagement with the world that lies at the heart of agency, distinguishing systems that exhibit genuine, self-directed behaviour from those that simply react to external forces in a predetermined manner [19, 25]. By continuously monitoring its surroundings and adjusting its decision-making accordingly, the agent demonstrates a level of situational awareness and adaptive capacity that transcends rigid, pre-programmed scripts [26–28]. Understanding this core principle lays the foundation for exploring the more sophisticated forms of autonomy as self-determination that emerge out of basic agency [29–32].

The enactive approach to cognition provides a useful lens for understanding the importance of environmental responsiveness in agency [33–35]. From this perspective, the agent's cognitive processes are not isolated from its physical embodiment and embeddedness within an ecological niche, but rather emerge through the ongoing, reciprocal interaction between the system and its surroundings [36–38]. Without

[1] We consider a more complex understanding of autonomy in the next section.

this engagement with the world, the system would lack the experiential foundation necessary to make meaningful, context-appropriate decisions. The capacity for environmental responsiveness is a hallmark of agentic experience, as it enables the system to fluidly adapt its behaviours in service of its goals [39].

The question of whether modern machine learning (ML) and artificial intelligence (AI) systems meet the fundamental criteria for agency—*interactivity, autonomy,* and *adaptability*—remains a matter of ongoing debate. While, for example, artificial neural networks (ANNs) can exhibit goal-directed behaviour in domains such as optimisation and reinforcement learning, their agency is inherently constrained by their design. These systems, though capable of performing highly specialised tasks, lack the autonomy or adaptability necessary to engage with the world in a dynamic, first-person manner. Unlike the bacterium discussed above, which continuously adapts based on its environment, ML systems operate within the rigid boundaries of pre-programmed objectives and data-driven algorithms. Their so-called 'goal-directedness', when viewed through the lens of Dennett's intentional stance, is more apparent than real, as they operate within the strict confines of pre-programmed objectives and data-driven algorithms.

The key issue here is that these systems are designed to optimise pre-specified outcomes. As Vermaas [40] points out, behaviours such as robot gestures—often interpreted as intentional actions—are merely the results of design and not a manifestation of intrinsic agency. Similarly, Papagni and Koeszegi [41] highlight the ethical and semantic limits of framing artificial 'agents' as intentional systems. This framework assumes that AI agents are inherently goal-directed, but in reality they are bound to follow the directives set by their programming and the data upon which they were trained. Thus, the agency attributed to modern AI is not the same as genuine, self-directed agency, but is rather an illusion or mimicking of agency created by goal-directed outputs tied to pre-programmed instructions—a phenomenon arguably illustrated by Searle [42] in his 'Chinese Room' argument.

Furthermore, as Floridi [43] argues, large language models such as ChatGPT demonstrate "agency without intelligence" or understanding, responding mechanically to prompts without the capacity to learn from real-time experiences [44]. These systems, reliant on vast amounts of static data, do not adapt dynamically to new contexts and cannot modify their behaviour autonomously. They are fundamentally reactive, unable to transcend the boundaries of their training [45]. This reflects a broader issue in contemporary AI discourse: the circular reasoning in the hypothesis that scaling these models—by adding data or computational power—somehow could enable the 'understanding' inherent to more advanced forms of agency [46–48]. The prevailing narrative, which frames AI's growing capabilities as a step toward autonomous intelligence, obscures the fact that these systems, for all their sophistication, remain static and heavily constrained by their initial design.

An enactive perspective highlights this gap further by strengthening the role of the body by seeing agents as embodied within a reciprocal engagement with an environmental niche [49, 50]. Without this, a system, like AI, cannot obtain real-time interaction, self-directed autonomy, and dynamic learning. While modern AI performs impressively in narrow domains, it remains far from meeting the philosophical and enactive standards of basic agency. Thus, while AI systems such as ChatGPT may exhibit increasingly sophisticated outputs, their limitations highlight the gap between the dynamic interactively situated nature of agency and its appearance. This distinction is crucial for shaping the future of AI development, particularly in terms of its ethical and philosophical implications, which we explore further below. We must recognise that AI is still far from achieving the sort of agency that is basic to human and biological systems.

We have defined agency as the capacity for environmentally responsive, goal-directed action, distinguishing it from mere mechanical causation or stimulus-response behaviours. However, whether enactive organisms can be said to possess consciousness—a scaled-up awareness beyond basic forms of agency—adds another layer to the agential dimension, complicating the relationship between the two concepts. At a basic level, agency can be observed in organisms such as bacteria, which navigate toward nutrients and away from harmful substances based on environmental cues, thereby engaging in purposeful behaviour without conscious awareness. On the one hand, this suggests that agency may be understood primarily as the capacity to interact with and adapt to the environment in a goal-directed manner, rather than being inherently tied to conscious experience. This leaves open the possibility for future forms of artificial basic agency that lack consciousness, which is important given the ongoing debate about whether artificial systems could ever achieve genuine consciousness [51, 52]. On the other hand, as we will see in the next section, conscious agency, as seen in humans and some non-human animals, involves higher-order processes such as self-reflection, decision-making, and long-term planning, involving enculturated practices, and values and norms underlying moral decision-making.

Ultimately, the relationship between agency and consciousness is nuanced. While conscious awareness may enrich the depth and flexibility of agency, the fundamental capacity for basic agency—the ability to interact with the environment in a goal-directed way—does not necessarily require consciousness. This distinction opens possibilities for understanding the agency of artificial systems, which may exhibit goal-directed behaviours without possessing consciousness. The next section will address the progression from basic to autonomous agency, focusing on self-directed goal setting, genuine choice, and the capacity to reflect on and adapt guiding principles, which are all key challenges for AI development.

2.2 Autonomous Agency

While basic agency involves goal-directed environmental responsiveness, full autonomous agency requires more sophisticated capacities for self-direction, authentic decision making, and critical reflection. Building on Formosa's and Ryan's work [53, 54], we can see autonomy as existing on a spectrum, starting from the most basic level of machine autonomy and extending through to the full autonomous agency characteristic of some mature humans. Basic machine autonomy can be understood as "the ability of a computer to follow a complex algorithm in response to environmental inputs, independently of real-time human input" [55, p. 149]. This basic form of autonomy allows us to differentiate between, for example, an autonomous vehicle (AV) that can navigate different routes independently and a remote-controlled car that requires constant human input to move. The AV has a level of basic machine autonomy that the simpler remote-controlled car lacks. But while there is a degree of machine autonomy in this example, it is not the same as the personal autonomous agency we see in mature humans. Full autonomy moves beyond mere environmental responsiveness to authentically shaping and directing an agent's engagement with the world.

Contemporary theories of personal autonomy typically distinguish between *authenticity* and *competency* conditions [56, 57]. Authenticity conditions require that the values and desires an agent acts on are genuinely their own, and not the result of manipulation, oppression, or undue external influence. Competency conditions specify that autonomous agents must possess certain skills and self-attitudes, such as the ability to critically reflect on their own values, adopt ends after self-reflection, and regard themselves as the bearers of dignity authorised to set and pursue their own ends among peers [58, 59]. These competencies also include appropriate self-attitudes, such as self-respect, self-love, self-trust, and self-esteem, which are needed to see oneself *as* a self-directing autonomous agent among others [54, 60, 61].

The first key requirement of autonomous agency is self-directed goal setting. While basic agents can pursue goals, autonomous agents must be able to generate, modify, and prioritise their own goals. This extends beyond merely having preferences or following pre-programmed objectives. Procedural theories of autonomy, such as Frankfurt's [62], argue that autonomy requires the capacity to form and act on second-order desires by not just wanting to do something (i.e. having preferences), but wanting to have that preference or desire. However, substantive theories of autonomy argue that these procedures alone are insufficient, as they struggle to address cases where agents reflectively endorse values, including second-order desires, which result from oppressive socialization [63, 64]. This emphasises that autonomous goal setting arguably requires both proper procedures, such as critical self-reflection, and appropriate substantive conditions, such as freedom from oppression.

The capacity for genuine choice is the second key requirement. Following Raz [65], autonomy requires not just selecting between options, but having access to an adequate range of valuable alternatives. Raz gives the example of a man stuck in a pit who has the choice between drinking a drop of water now or later. While the man in the pit has choices, they are so meagre and limited that we would not see him as a genuinely autonomous agent directing his own life. Current AI systems can weigh options against pre-programmed criteria, but this differs from genuine autonomous choice-making that involves selecting from a range of meaningful options and situating decisions within a framework of critically self-endorsed values.

The third element involves the capacity for critical reflection on and potential modification of guiding values and principles. This requires a degree of self-consciousness through awareness of yourself as a distinct agent capable of examining and revising the normative framework within which you make choices [60]. Drawing on Watson [66], this involves the capacity for reflective endorsement of your values and choices, which requires both procedural abilities for critical reflection and substantive conditions that enable authentic value formation free from oppression. Current AI systems, while able to process feedback and adjust behaviours, lack this capacity for authentic revision of their fundamental values and normative frameworks. Indeed, in many ways current AI systems resemble agents formed under conditions of strong oppressive socialization, since their "values" and decision-making frameworks are externally imposed on them through training, data, and feedback which (often) encodes existing human biases, while the systems themselves lack the capability to critically revise these imposed normative structures.

While contemporary AI systems can make complex decisions and demonstrate sophisticated environmental responsiveness, including formulating linguistic responses that make sense and respond dynamically to human input [46], they lack the authenticity and competency conditions required for genuine autonomous agency. Mapping them on to a spectrum from basic machine autonomy through to full autonomous agency, current AI systems, while capable of sophisticated independent operation, remain closer to the basic machine autonomy end of this spectrum. This aligns with our earlier analysis of the limitations of AI systems in achieving even basic agency, since their apparent goal-directedness remains deeply constrained by their initial programming and training. AI systems also lack the elements needed for full autonomous agency, including authentic self-direction, choices grounded in self-endorsed values, appropriate self-attitudes, a meaningful range of options, and the capacity for critical self-reflection. For example, while a LLM might use linguistic phrases such as "I respect myself," it cannot genuinely develop self-respect as it lacks the capacity for authentic self-understanding and self-valuing that these attitudes require. This has important implications for how we think about AI systems, particularly if we regard full autonomous agency as deserving of moral respect and ethical consideration.

3 Moral Agency and Moral Patients

Having examined variations of both *basic* and *autonomous* agency, we now turn to *moral* agency. Moral agency is often thought to apply to only those agents that can be justifiably held morally responsible for their actions [67, 68]. Moral agency normatively qualifies basic agency by holding that the behaviour of moral agents can and ought to be guided by moral requirements, such as moral duties and obligations, that they can be held responsible for meeting [69]. When a moral agent fails to guide their behaviour according to moral requirements by violating a duty or obligation, they are deserving of moral blame, unless various excusing or exempting conditions apply [70]. Clearly, not all agents are moral agents. For example, a bear stealing food from campers would not be deserving of moral blame for stealing. This is because, even though the bear is a conscious agent that engages in purposeful behaviour and responds to environmental cues, it is (at least standardly) held not to be a moral agent that is subject to moral requirements.

Following Himma [69], according to the standard view, there are two jointly necessary and sufficient features of moral agency, which are both necessarily realised through consciousness. The first is the capacity for *rational deliberation*. Rationality is seen as necessary for moral agency as it contributes toward, along with autonomy, grounding an agent's capacity to freely reason and deliberate. It is a moral agent's capacity to make free decisions via rational deliberation that contributes toward their being justifiably held responsible for their actions. This condition aligns with the above account of autonomous agency, which requires agents to have the capacity to make a genuine choice that is self-directed, and earlier considerations about the situational constraints and environmental conditions that agents face. The second is an epistemic condition that an agent can only be held morally responsible for their acts if they have the capacity to identify moral requirements by being adequately familiar with moral concepts and the application of basic moral principles; for example, *knowing* that it is wrong to steal. This epistemic condition also aligns well with autonomous agency requiring the conscious capacities to critically reflect on, and adjust, one's own normative commitments and actions based on self-reflection.

According to the standard view, the capacities that are considered necessary and sufficient for moral agency appear to implicitly require that moral agents are conscious, since moral *deliberation* seems to require consciousness [69]. While the standard view sees consciousness as necessary for moral agency, it does not follow that *all* conscious agents are moral agents (e.g., such as the bear in our above example), or that *all* agents are conscious (e.g., such as the bacterium example from Sect. 2.1), but only that consciousness is required for moral deliberation and thus also for *moral* agency. However, we shall explore in the next section whether artificial agents can challenge the assumption that moral deliberation requires consciousness.

While moral agency captures an agent's capacity to be a source of moral action, moral *patiency* refers to entities that can be the subject of moral obligations or duties [13, 71]. To be a moral agent or moral patient is to have moral standing. According to the standard view of moral agency, *only* moral agents are moral patients and *vice versa* [13]. The basis of this view is that only moral agents have moral standing in virtue of their intrinsic, rather than instrumental, value (or innate dignity), realised via the distinctly human capacity of personhood or full autonomous agency. A view of this form is commonly ascribed to Kant, who holds that we can only have *direct* moral duties to other rational agents, whereas we can only have *indirect* duties to non-human animals in virtue of the impacts our treatment of them could have on ourselves and other rational agents ([72]; for critical attempts to expand Kantian views to other forms of agency see, e.g., [73–75]).

However, work in animal and environmental ethics has complicated the relationship between moral agency and moral patiency, and as a result has opened the door to a consideration of moral patiency as distinct from moral agency (e.g., [71, 73, 76–78]). The key insight to draw here from what Floridi and Sanders [13] refer to as the "non-standard view" is that, while all moral agents are also moral patients, it is not the case that all moral patients are moral agents. The basis of this extension is that certain entities that are not moral agents nonetheless can experience morally significant impacts on their well-being or have morally significant interests that can give rise to direct moral obligations or duties, even if they cannot engage in moral deliberation themselves. The class of moral patients is (among other cases) typically expanded to include non-human animals, without including them as moral agents. Thus, while the bear from our earlier example cannot be held morally responsible for its actions since it is not a moral agent, moral agents can be held responsible for actions directed towards the bear as it is a moral patient. For example, it would be immoral for a moral agent to torture the bear for fun, given that the bear can feel pain and has valuable interests, since such actions fail to treat the bear properly given its status as a moral patient. Given the apparent agency and machine autonomy of AI systems, the question arises, which we explore in the next section, whether we should expand the realm of moral standing to include them.

4 The AI Challenge to the Link Between Agency and Patiency

4.1 Artificial Moral Agency

What has resulted from this expansion of moral patiency is a deep gulf between those eligible for moral patiency, such as bears and other non-human animals, and moral agency, which is located primarily or even exclusively in (some) humans. Floridi and Sanders [13] argue that this gulf has resulted from a deep concern over the character of moral patiency that has not been similarly extended to moral agency. They argue that we ought to close this gap, not by adopting the standard narrow view of moral agency that restricts all moral patients to moral agents (understood as those capable of moral deliberation), but rather by expanding out a conception of moral agency that includes all moral patients, such as non-human animals, and even distributed moral systems, such as ecological, social and legal entities, and thus not only those systems and entities that are capable of moral deliberation. They further argue that while this may come at the intuitive cost of including artificial AI agents *as* moral agents—or Artificial Moral Agents (AMAs) for short—it is a cost some consider worth paying (see, e.g., [79]). Is this expansion plausible?

To understand whether AI systems could challenge the relationship between moral agency and patiency, we need to first outline different levels of potential AMAs. Following Moor [80] and building on Formosa and Ryan's contribution [53], we can differentiate between four increasingly sophisticated levels of AMAs. The first two levels, which Moor calls "ethical impact agents" and "implicit ethical agents," both have the capacity to cause morally relevant outcomes, which is true of many artifacts, but are not moral agents in any interesting or substantive sense. Formosa and Ryan [53] use the example of a toaster. Imagine a dumb toaster that burns your hand when it is too hot, and a slightly less dumb toaster that has been designed with implicit ethical principles in mind such as a light that warns the user when it is too hot to touch. While both toasters could produce ethically salient outcomes (i.e., they could cause harm to humans), and the latter toaster is designed with ethical outcomes in mind (i.e., to minimise harm), neither is a moral agent. Neither toaster acts in a *self-directed* and *adaptable* way nor acts as the result of moral deliberation; in other words, the toaster does not decide to turn on the light *because* it knows that it is a morally good thing to do.

The more interesting cases are Moor's [80, 81] third and fourth levels: "explicit ethical agents" and "full ethical agents". Explicit ethical agents appear to act *from* ethical principles rather than merely being designed to respond *according to* them. An explicit ethical agent toaster would, for example, turn its light on when it reaches a particularly dangerous temperature, not because it is designed to minimise harm, but rather *because* its actions are a product of moral deliberation about the moral obligation to warn users of possible harm. But such a hypothetical toaster is still very limited in what it can do. It can only make toast and turn a light on or off. Formosa and Ryan [53] see the (potential) moral agency of explicit ethical agents as emerging across a continuum of cases, with only entities capable of explicit ethical agency across a sufficiently varied range of novel cases potentially counting as moral agents in this limited sense. For instance, a hypothetical advanced AMA system that, while non-conscious, could analyse complex situations, identify relevant moral considerations, weigh competing ethical principles, and make ethically grounded decisions across a rich range of novel cases might count as some form of (at least limited) moral agent. This might also require, as discussed in Sect. 2, some form of embodiment or at least the ability to sense and respond to its environment. Finally, "full ethical agents" add "consciousness, intentionality, and free will" [81] to these capabilities, thereby achieving the kind of complete moral agency characteristic of mature humans. However, since this final level requires the sort of conscious AI that many doubt is possible, we do not consider it any further here.

The possibility of such cases raises a crucial question: does what *appears to be* moral deliberation in (hypothetical) advanced artificial explicit ethical agents really count as *genuine* moral deliberation? As we saw in our earlier analysis of basic agency and autonomy, there are reasons for skepticism here. Just as current AI systems can appear to be agent-like or autonomous without really being so, these more advanced systems might exhibit the appearance of moral deliberation without genuinely morally deliberating. Their apparent ethical decision-making may simply be sophisticated pattern matching rather than authentic moral reflection. In any case, Wallach and Allen [82] argue that the most promising approach in the machine ethics literature to make sense of non-conscious artificial moral agency is via a hybrid approach, which synthesises both top-down and bottom-up approaches to moral deliberation (see also [83]). A hybrid approach would attempt to supervise this training process by filtering it through normative ethical heuristics. Thus, while non-conscious, it seems plausible that some explicit ethical agents that exhibit a hybrid approach to moral decision-making are engaged, or at least appear to be engaged, in a kind of deliberative process that may minimally ground some form of (albeit limited) moral agency.

This potential raises an important question: *if* AI systems could potentially exhibit at least a limited form of moral agency without consciousness, could these entities challenge the standard coupling of moral agency and moral patiency? To explore this possibility, we need to examine more carefully whether and how AI systems could qualify as moral patients. Of course, those who claim that moral agency *in any form* requires consciousness will not accept this claim. Nonetheless, we think that it is at least interesting to consider that there *might* be some forms of *limited* moral agency that could exist without consciousness, which is sufficient to motivate our discussion of this possibility in the next section.

4.2 Artificial Moral Patients

We have so far considered both the standard view of moral agency, which sees *all* and *only* moral agents as moral patients, and the non-standard view, which extends moral patiency beyond moral agents to include non-human animals, distributed moral systems, and even AMAs on some views. Despite their differences, both views seem committed to the claim that *all* moral agents are moral patients. Might the case of possible artificial moral agents lead us to question this claim? As we saw in Sect. 2.1, some forms of basic agency may be possible without consciousness, and in Sect. 4.1 we saw that some form of limited moral agency might be possible for non-conscious artificial systems through hybrid approaches to moral decision-making. While this view can be readily rejected by those who require consciousness for moral deliberation [69], it is still worth exploring whether potentially granting some degree of limited moral agency to artificial systems could lead us to question the conceptual claim that moral agency implies moral patiency.

The claim that moral agency implies moral patiency is commonly made. For example, Himma writes that "moral agents are usually, if not always, moral patients" [69, p. 21], while Floridi and Sanders [13, p. 377] argue that the category of moral agents that are not moral patients is "utterly unrealistic" as it would require a "pure agent" that is "in principle 'unaffectable'". However, we suggest that AMAs might present exactly this kind of case—entities that are capable of (at least the appearance of) moral deliberation and (limited) moral agency without qualifying as moral patients. Such entities could undertake some form of moral deliberation, but not be the subject of moral duties.

The plausibility of this suggestion is based on a proposed necessary link between consciousness and moral patiency.[2] Following Birch's [84] appeal to sentience, moral patiency may require the capacity to register phenomenally valenced experiences—that is, to experience the promotion or frustration of one's interests through subjective conscious awareness. Even those who lack the capacity for valenced

[2] We recognise that this suggestion requires much more defence than we can give it here, and that it may have problems of its own when we consider the moral patiency of non-conscious entities such as ecosystems. Nonetheless, we think this suggestion is both plausible and interesting in what it might imply about artificial systems.

experience could nonetheless *register* their interests being promoted or frustrated without the need to appeal to the valence of experience via explicit rational endorsement [84, 85]. For example, a philosophical Vulcan that has no valenced experience can still consciously register that its interests have been frustrated. In either case, it is the capacity to *register*, through *conscious experience*, one's interests that gives rise to moral patiency on this view. As such, consciousness appears necessary for moral patiency, even if consciousness is not necessary for moral agency.[3] This could decouple moral patiency from moral agency, and thus even if an artificial system lacking consciousness could count as a limited moral agent, it need not follow that it is therefore a moral patient as well.

The tendency to assume that the claim to patiency regarding technological artifacts, such as AMAs, can only be supported by appealing to their agency, might stem from an analogical error that is often evoked between machine and animal ethics (see [86, 87]). This analogy fails precisely because artificial systems, unlike many non-human animals, lack consciousness. And it is precisely higher-ordered features, such as consciousness, that humans share with (at least many) non-human animals that underwrites much of the push to extend moral standing to non-human animals. This suggests that attempts to push the case for the extension of moral patiency to AMAs via an analogous appeal to the extension of moral patiency to non-human animals is likely to be flawed [79, 86, 87].

Nonetheless, some might argue that we should, on precautionary grounds, be wary of dismissing appeals for extending moral patiency to artificial systems that *appear* to behave in ways that would normally ground moral standing, such as engaging in what looks like moral deliberation that impacts the world [88, 89]. Others suggest we should treat AMAs ethically for instrumental reasons, even if they are not genuine moral patients [87, 90, 91]. These approaches are clearly worthy of consideration, and it may be that we could have instrumental or pragmatic reasons for treating AMAs *as if* they are moral patients, but that does not impact the point we are making here that they would nonetheless not really *be* moral patients if they lacked consciousness.

5 Conclusion

This paper has developed a systematic analysis of agency in its various forms, including basic, autonomous, and moral agency, and critically examined the extent to which AI systems might instantiate each form of agency. We have shown that while current AI systems are highly sophisticated, they lack genuine agency and autonomy because: they operate within rigid boundaries of pre-programmed objectives rather than exhibiting true goal-directed behaviour within their environment; they cannot authentically shape their engagement with the world; and they lack the critical self-reflection and autonomy competencies required for full autonomy.

[3] Keep in mind that we are merely exploring the implications of rejecting the claim that moral agency requires consciousness, rather than assuming or arguing for its rejection here.

Nonetheless, we do not rule out the possibility of future systems that could achieve a limited form of artificial moral agency without consciousness through hybrid approaches to ethical decision-making. This leads us to suggest, by appealing to the necessity of consciousness for moral patiency, that such non-conscious AMAs might represent a case that challenges traditional assumptions about the necessary connection between moral agency and moral patiency. Our analysis has implications for both our theoretical understanding of agency and our practical approach to AI development. It suggests we need more nuanced frameworks for understanding different types and degrees of artificial agency, while challenging us to rethink assumptions about moral standing. As AI systems become more sophisticated, these conceptual distinctions will become increasingly important for guiding their ethical development and deployment.

Competing Interests The authors have no conflicts of interest to declare that are relevant to the content of this chapter.

References

1. Fulda FC (2017) Natural agency: the case of bacterial cognition. J Am Philos Assoc 3(1):69–90
2. Fulda FC (2023) Agential autonomy and biological individuality. Evol Dev 25(6):353–370
3. Garland T Jr, Downs CJ, Ives AR (2022) Trade-offs (and constraints) in organismal biology. Physiol Biochem Zool 95(1):82–112
4. Küpers WM (2024) Integrating embodied practice and transformational wisdom for sustainable organization and leadership. Routledge, New York
5. Oyama S (2000) The ontogeny of information. Duke University Press, Durham
6. De Beauvoir S (1947) The ethics of ambiguity (trans: Frechtman B). Citadel Press, New York
7. Corris A (2022) An enactive-developmental systems framing of cognizing systems. Biol Philos 37(4):32
8. Jaeger J (2023) The fourth perspective: evolution and organismal agency. In: Mossio M (ed) Organization in biology. History, philosophy and theory of the life sciences, vol 33. Springer, Cham, pp 159–186. https://doi.org/10.1007/978-3-031-38968-9_8
9. Jaeger J (2024) Ontogenesis, organisation, and organismal agency. In: Švorcová J (ed) Organismal agency. Biosemiotics, vol 28. Springer, Cham, pp 165–190. https://doi.org/10.1007/978-3-031-53626-7_10
10. Anscombe E (1957) Intention. Harvard University Press, Cambridge, MA
11. Segundo-Ortin M (2020) Agency from a radical embodied standpoint. Front Psychol 11:1319
12. Hodges BH (2023) Values define agency: ecological and enactive perspectives reconsidered. Adapt Behav 31(6):559–576
13. Floridi L, Sanders JW (2004) On the morality of artificial agents. Mind Mach 14(3):349–379. https://doi.org/10.1023/B:MIND.0000035461.63578.9d
14. Ryan RM, Deci EL (2024) Self-determination theory. In: Maggino F (ed) Encyclopedia of quality of life and Well-being research. Springer, Cham, pp 6229–6235. https://doi.org/10.1007/978-3-031-17299-1_2630
15. Weichold M (2024) The autonomous individual. Taylor & Francis, New York
16. Beer RD, Di Paolo EA (2023) The theoretical foundations of enaction. Biosystems 223:104823
17. Heylighen F (2023) The meaning and origin of goal-directedness. Biol J Linn Soc 139(4):370–387
18. Dennett DC (1987) The intentional stance. MIT Press, Cambridge, MA
19. Sharov A, Tønnessen M (2022) Semiotic agency: science beyond mechanism. Springer, Cham

20. Huffermann JD (2023) The primacy of knowing-how: cognition, know-how and an enactive action first epistemology. Doctoral thesis, Universidade Federal do Rio Grande do Sul. https://lume.ufrgs.br/bitstream/handle/10183/259197/001171149.pdf?sequence=1
21. Bongard J, Levin M (2021) Living things are not (20th century) machines. Front Ecol Evol 9:650726
22. Jaeger J, Riedl A, Djedovic A, Vervaeke J, Walsh D (2024) Naturalizing relevance realization: why agency and cognition are fundamentally not computational. Front Psychol 15:1362658
23. Fuchs T (2020) The circularity of the embodied mind. Front Psychol 11:1707
24. Rouse J (2023) Social practices as biological niche construction. University of Chicago Press, Chicago
25. Siqueiros-García JM, Manuel-Navarrete D, Eakin H, Mojica L, Charli-Joseph L, Pérez-Belmont P, Ruizpalacios B (2022) Sense of agency, affectivity and social-ecological degradation. Front Psychol 13:911092
26. Carvalho EMD, Rolla G (2020) An enactive-ecological approach to information and uncertainty. Front Psychol 11:588
27. Gallagher S (2023) Embodied and enactive approaches to cognition. Cambridge University Press, Cambridge
28. Sepúlveda-Pedro MA (2023) Enactive cognition in place. Palgrave Macmillan, Cham
29. Fuchs T (2017) Ecology of the brain. Oxford University Press, Oxford
30. Ryan JKJ, Gallagher S (2020) Between ecological psychology and enactivism. Front Psychol 11:1147
31. Favela LH (2023) The ecological brain. Routledge, New York
32. Heras-Escribano M (2023) Ecological psychology, enaction, and the quest for an embodied and situated cognitive science. In: Casper MO, Artese GF (eds) Situated cognition research: methodological foundations. Springer, Cham, pp 83–102. https://doi.org/10.1007/978-3-031-39744-8_6
33. Varela FJ, Thompson E, Rosch E (2017) The embodied mind. MIT Press, Cambridge, MA
34. Gallagher S (2023) Surprise! Why enactivism and predictive processing are parting ways. Possibility Stud Soc 1(3):269–278
35. Barrett FN (2024) Experience and nature in pragmatism and enactive theory. Phenomenol Cogn Sci 24:1–23
36. Gallagher S (2020) Action and interaction. Oxford University Press, Oxford
37. Hutto D, Gallagher S, Ilundáin-Agurruza J, Hipólito I (2020) Culture in mind—an enactivist account. In: Kirmayer LJ, Kitayama S, Worthman CM, Lemelson R, Cummings CA (eds) Culture, mind, and brain. Cambridge University Press, Cambridge, pp 163–187
38. Gahrn-Andersen R (2023) Code biology and enactivism. Bringing adaptors to basic minds. In: Aldini A (ed) Software engineering and formal methods. SEFM 2023 collocated workshops. SEFM 2023. Lecture notes in computer science, vol 14568. Springer, Cham, pp 114–129
39. Rolla G, Figueiredo N (2023) Bringing forth a world, literally. Phenomenol Cogn Sci 22(4):931–953
40. Vermaas P (2024) Describing robot gestures by design and agency. In: Breyer T, Gerner AM, Grouls N, Schick JFM (eds) Diachronic perspectives on embodiment and technology: gestures and artefacts. Springer, Cham, pp 83–95
41. Papagni G, Koeszegi S (2021) A pragmatic approach to the intentional stance semantic, empirical and ethical considerations for the design of artificial agents. Mind Mach 31(4):505–534
42. Searle J (2009) Chinese room argument. Scholarpedia 4(8):3100
43. Floridi L (2023) AI as agency without intelligence. Philos Technol 36(1):15
44. Pütz O, Esposito E (2024) Performance without understanding. Discourse Commun 18(6):859–868
45. Titus LM (2024) Does ChatGPT have semantic understanding? A problem with the statistics-of-occurrence strategy. Cogn Syst Res 83:101174
46. Hagendorff T, Fabi S, Kosinski M (2023) Human-like intuitive behavior and reasoning biases emerged in large language models but disappeared in ChatGPT. Nat Comput Sci 3(10):833–838
47. Surden H (2023) ChatGPT, AI large language models, and law. Fordham L Rev 92:1941

48. Yang J, Jin H, Tang R, Han X, Feng Q, Jiang H, Yin B, Hu X (2024) Harnessing the power of LLMs in practice: a survey on ChatGPT and beyond. ACM Trans Knowl Discov Data 18(6):1–32
49. Gonçalves DA, Caceffo R, Valente JA, Bonacin R, dos Reis JC, Baranauskas MCC (2024) Enactive interaction in technology-based learning environments. Educ Technol Soc 27(2):87 113
50. Di Paolo EA, Lawler D, Vaccari AP (2023) Toward an enactive conception of productive practices. Philos Technol 36(2):31
51. Torrance S (2008) Ethics and consciousness in artificial agents. AI & Soc 22(4):495–521
52. Schneider S, Turner P (2017) Is anyone home? A way to find out if AI has become self-aware. Scientific American
53. Formosa P, Ryan M (2021) Making moral machines. AI Soc 36(3):839–851
54. Formosa P (2021) Robot autonomy vs. human autonomy. Mind Mach 31(4):595–616
55. Etzioni A, Etzioni O (2016) AI assisted ethics. Ethics Inf Technol 18(2):149–156
56. Christman J (2009) The politics of persons. Cambridge University Press, Cambridge
57. Susser D, Roessler B, Nissenbaum H (2019) Technology, autonomy, and manipulation. Internet Policy Rev 8(2):1–22
58. Meyers DT (1987) Personal autonomy and the paradox of feminine socialization. J Philos 84(11):619–628
59. Formosa P (2013) Kant's conception of personal autonomy. J Soc Philos 44(3):193–212
60. Benson P (1994) Free agency and self-worth. J Philos 91(12):650–658
61. Mackenzie C (2008) Relational autonomy, normative authority and perfectionism. J Soc Philos 39(4):512–533
62. Frankfurt HG (1971) Freedom of the will and the concept of a person. J Philos 68(1):5–20
63. Benson P (1991) Autonomy and oppressive socialization. Soc Theory Pract 16(3):385–408
64. Mackenzie C, Stoljar N (2000) Relational autonomy. Oxford University Press, Oxford
65. Raz J (1986) The morality of freedom. Clarendon Press, Oxford
66. Watson G (1975) Free agency. J Philos 72(8):205–220
67. Sparrow R (2007) Killer robots. J Appl Philos 24(1):62–77
68. Sparrow R (2011) Can machines be people? Reflections on the Turing triage test. In: Lin P, Abney K, Bekey GA (eds) Robot ethics. MIT Press, Cambridge, MA, pp 301–315
69. Himma KE (2009) Artificial agency, consciousness, and the criteria for moral agency: what properties must an artificial agent have to be a moral agent? Ethics Inf Technol 11(1):19–29. https://doi.org/10.1007/s10676-008-9167-5
70. Formosa P (2006) Moral responsibility for banal evil. J Soc Philos 37(4):501–520. https://doi.org/10.1111/j.1467-9833.2006.00354.x
71. Regan T (2004) The case for animal rights. University of California Press, Oakland
72. Kant I (1996) The metaphysics of morals. In: Practical philosophy. Cambridge University Press, Cambridge, pp 353–604
73. Korsgaard CM (2018) A Kantian case for our obligations to the other animals. In: Fellow creatures. Oxford University Press, Oxford
74. Wood A (1998) Kant on duties regarding nonrational nature. Proc Aristot Soc Suppl 72:189–210
75. Formosa P (2017) Kantian ethics, dignity and perfection. Cambridge University Press, Cambridge
76. White L (1967) The historical roots of our ecologic crisis. Science 155(3767):1203–1207
77. Singer P (ed) (1977) Animal liberation. Avon Books, New York
78. Kuhse H, Singer P (2002) Individuals, humans and persons: the issue of moral status. In: Unsanctifying human life. Blackwell, Oxford
79. Gunkel DJ (2014) A vindication of the rights of machines. Philos Technol 27(1):113–132
80. Moor J (2009) Four kinds of ethical robots. Philos Now 72:12–14
81. Moor JH (2006) The nature, importance, and difficulty of machine ethics. IEEE Intell Syst 21(4):18–21
82. Wallach W, Allen C (2009) Moral machines. Oxford University Press, Oxford

83. Allen C, Smit I, Wallach W (2005) Artificial morality. Ethics Inf Technol 7(3):149–155. https://doi.org/10.1007/s10676-006-0004-4
84. Birch J (2024) The edge of sentience. Oxford University Press, Oxford
85. Johnson DG (2006) Computer systems: moral entities but not moral agents. Ethics Inf Technol 8(4):195–204
86. Johnson DG, Verdicchio M (2018) Why robots should not be treated like animals. Ethics Inf Technol 20(4):291–301
87. Levy D (2009) The ethical treatment of artificially conscious robots. Int J Soc Robot 1(3):209–216
88. Danaher J (2020) Welcoming robots into the moral circle. Sci Eng Ethics 26(4):2023–2049. https://doi.org/10.1007/s11948-019-00119-x
89. Danaher J (2023) Moral uncertainty and our relationships with unknown minds. Camb Q Healthc Ethics 32(4):482–495. https://doi.org/10.1017/S0963180123000191
90. Coeckelbergh M (2021) Should we treat teddy bear 2.0 as a Kantian dog? Four arguments for the indirect moral standing of personal social robots, with implications for thinking about animals and humans. Mind Mach 31(3):337–360
91. Friedman C (2023) Granting negative rights to humanoid robots. Front Artif Intell Appl 366:145–154. https://doi.org/10.3233/FAIA220613

Paul Formosa is Professor of Philosophy and Research Training Director in the School of Humanities at Macquarie University. He previously served as Head of the Department of Philosophy, founding co-Director of the Macquarie University Ethics and Agency Research Centre, and Director of the Centre for Agency, Values and Ethics (CAVE). A former Macquarie University Research Fellow and Australian Research Council (ARC) DECRA Fellow, Professor Formosa has been a Chief Investigator on three ARC Discovery Projects and various industry-funded research projects. He has published extensively in moral and political philosophy, including work on Kantian ethics, moral evil, moral psychology, human dignity, autonomy, and the philosophy of technology (with a particular focus on the ethics and politics of Artificial Intelligence (AI), social media, and video games). Professor Formosa has produced over 100 research outputs, and his work is highly cited.

Inês Hipólito is an internationally recognised philosopher of cognitive science and artificial intelligence, and currently an Assistant Professor at Macquarie University. Her research bridges embodied cognition, complex systems, and the Free Energy Principle to advance non-representational, context-sensitive AI. Prior to her current role, she was a lecturer at Humboldt-Universität zu Berlin's Berlin School of Mind and Brain and is co-PI on the €1M Berlin University Alliance Neuro-Urbanism project. She is co-founder and former President of the International Society for the Philosophy of the Sciences of the Mind, and serves on the Australasian Association of Philosophy's Media Committee. With over 50 peer-reviewed publications, eight special issues, and keynote appearances at major events including Your Brain on AI at the Sydney Opera House and the Royal College of Psychiatry, she is an influential voice in philosophy of mind and active inference.

Thomas Montefiore is an early career researcher in the School of Humanities at Macquarie University. Dr Montefiore is a Postdoctoral Research Fellow on an Australian Research Council (ARC) Discovery Project and a member of the Macquarie University Ethics and Agency Research Centre. Thomas' work primarily focuses on moral philosophy and the applied ethics of emerging technologies—such as video games, virtual reality, synthetic meat, death technology and social robots.

Part II

Human and Machinic Autonomy. Ethical Implications

Automated Gatekeepers: How Recommender Systems Shape and Constrain Autonomy

4

Hugo Cossette-Lefebvre and Natalie Stoljar

Abstract

This chapter explores the impact of recommender systems on autonomy. We draw on feminist relational approaches to argue that recommender systems can threaten autonomy in four key ways. (i) Recommender systems limit people's exposure to options, creating tunnel vision. (ii) They undermine the imaginative capacities crucial to the critical reflection required for autonomy and constrain the options people take to be feasible for them. (iii) They can generate a "lens" through which people experience the world, reinforcing oppressive stereotypes and negatively affecting self-trust or self-respect. (iv) They create echo chambers that generate an epistemological problem for autonomy by privileging false beliefs. After identifying these threats, we argue that recommender systems are not inherently autonomy-undermining and could be designed to promote autonomy. To achieve this, we propose three design goals: (i) ensuring exposure to diverse content to prevent tunnel vision; (ii) safeguarding self-respect and avoiding the exploitation of individual vulnerabilities; and (iii) generally promoting truthfulness.

Keywords

Recommender systems · Relational autonomy · Echo chambers · Deliberative skills · Self-respect · Social hierarchies

H. Cossette-Lefebvre (✉) · N. Stoljar
McGill University, Montreal, QC, Canada
e-mail: hugo.cossette-lefebvre@mail.mcgill.ca

M. Anzalone et al. (eds.), *Reconfiguring Human Autonomy*, Integrated Science 40,
https://doi.org/10.1007/978-3-032-14537-6_4

1 Introduction

In this chapter, we examine the ways in which recommender systems (and relatedly, filter bubbles or echo chambers) affect personal autonomy. We draw on feminist relational conceptions of autonomy to argue that exposure to recommender systems can negatively affect autonomy and hence wrong users. Recommender systems are problematic for autonomy for two broad reasons: First, they function to unacceptably narrow the scope of users' options, which both undermines autonomy directly *and* hampers the skills of critical reflection required for autonomy. Second, recommender systems push certain *contents* in ways that reinforce dominant narratives that undermine autonomy. For instance, content promoting being a "traditional wife", or "tradwife" for short, is currently ubiquitous. It advocates for the rejection of feminism and encourages women to stay home, keep house, have children, and generally to be subservient to their husbands [1]. This content is reinforced by masculinist influencers, like Andrew Tate, who openly promote misogyny and essentialist understandings of gender [2].

Section 2 of the chapter elaborates on recommender systems and how they operate. We explain the link between recommender systems and echo chambers. We also briefly review the ethical concerns that philosophers have already identified with respect to these two common aspects of internet and social media use. The main argument of the chapter is presented in section three. It identifies the dimensions of relational autonomy, and then proposes four reasons that recommender systems and echo chambers, as they currently operate, risk undermining autonomy. (i) Recommender systems present people with options that are very similar to already-existing preferences, and thereby limit people's exposure to a wide variety of options. They can create a kind of "tunnel vision" regarding options which on relational accounts is directly incompatible with autonomy. (ii) Recommender systems undermine the imaginative capacities that are crucial to the critical reflection required for autonomy, and unacceptably constrain the options that people take to be feasible for them. (iii) Recommender systems generate a "lens" through which people experience the world and/or reinforce the already-existing lens. When the content that is pushed by recommender systems reinforces dominant, oppressive stereotypes—such as when it normalizes sexist stereotypes—this creates a lens that is incompatible with the self-respect that is necessary for autonomy. (iv) Recommender systems can create echo chambers which generate an epistemological problem for autonomy by privileging false beliefs. When this is the case, autonomy is undermined for people who receive their information exclusively through these echo chambers. Once having outlined the potentially autonomy-undermining features of recommender systems, we turn in the fourth section to a brief discussion of how recommender systems might be modified to avoid these consequences.

2 Recommender Systems, Filter Bubbles, and Echo Chambers

Contemporary artificial intelligence systems play a structuring role in curating what content users are exposed to on diverse apps and websites. They are ubiquitous in our day-to-day lives and they delimit the options that are presented to us online. Broadly, recommender systems are "software tools and techniques that provide suggestions for available items that are most likely of interest to a particular user" [3, 4]. They are, consequently, necessary and immensely valuable for navigating the virtually infinite amount of online content that is available. A Google search that succeeds in presenting quickly what will be most relevant to a user allows her to achieve her goals quickly and efficiently, much more so than wading through virtually endless content by herself. Yet, recommender systems can also be problematic because they are not neutral tools: they can promote problematic content (such as falsehoods and misinformation) and create a form of tunnel vision by filtering the content users are exposed to.

Recommender systems powered by machine-learning techniques collect, curate and act upon vast amounts of personal data to infer the contents a user is most likely to interact with based on the user's past behavior and the data of other users [5–7]. Recommender systems infer a person's preferences by tracing connections between who the user is (or, rather, who the user is believed to be by the system), how the user behaves in the system, and the behavior of similar others. A university-educated woman in her mid-twenties who likes to listen to pop music will receive different recommendations than a high-school-educated man in his thirties who likes to crochet. Results in a search engine or a social media feed will differ for every user; two people with the same friends or who follow the same people or the same pages on a form of social media will be exposed to different contents depending on their past interactions with the system, and the information the system has or infers about them.

More precisely, we can distinguish between two types of filtering used by recommender systems. First, *content-based filtering* analyzes the associations between a user's past behaviors and the description of new objects to match individual observed preferences and new content. Second, *social information filtering,* or *collaborative filtering* aims to automate the process of word of mouth. This second process recommends items to a user based on the tastes that are assigned to them in the light of data from people with similar behavior [5, 8, 9].[1] These types of filtering are, of course, compatible: it is possible to consider both individual behavior and complement it with social filtering. Recommender systems also aim to be adaptive, interactive processes. They aim to adapt dynamically to changes in both individual and population behavior. This allows them to identify "trending" content, for example.

Recommender systems have been subject to ethical criticism on a number of grounds. Privacy concerns arise from the information that is collected and the

[1] Note that this collaborative filtering can track categories that are completely *ad hoc* and need not be endorsed by the user or even socially meaningful. These categories can include "dog owner," "recently bought a novelty sweater," or "likes curly fries."

inferences that are made about individuals and groups on the basis of that information [6]. There is also a risk of third party coercion and manipulation: those who control the recommender systems can nudge or target vulnerable users in ways that constitute exploitation, or favour certain types of content that will then be pushed on users [4, 5, 10]. Although these are urgent and real ethical issues, we focus in this chapter specifically on how recommender system filtering techniques can threaten, or promote, autonomy understood relationally.

However, before outlining the main argument, we note two prima facie problems that follow from existing filtering techniques. Recommender systems, by their very design, limit the number of options to which users are exposed. As mentioned above, this is not necessarily a problem since it is valuable to be able to navigate an endless stream of data and information efficiently. Nevertheless, recommender systems are designed to promote specific goals, in particular to maximize user engagement on platforms such as YouTube, X, TikTok, etc. This leads to the first issue: although maximizing user engagement is not necessarily grounded in malicious intent, it can have bad consequences. As Alfano and colleagues [11] (see also [12]) have shown, YouTube's recommender system has the tendency to promote extremist content, conspiracy theories, and fake news, because such videos are especially likely to capture and keep users' attention. If a person relies heavily on social media as an information source, the information they will use to guide their own actions, as well as to form their beliefs and preferences, is likely to be biased or even outright false. In other words, the purposes for which recommender systems are used—such as to maximize engagement—can be inconsistent with what is actually valuable for individuals or communities—such as access to reliable information.

The second issue is that personalization may create filter bubbles or echo chambers which prevent people from seeing content that would otherwise interest them or allow them to avoid facts and opinions with which they disagree.[2] By relying heavily on past behavior and similarities with others, recommender systems can fail to accurately capture the actual preferences of a person and how persons can change over time. By failing to expose them to content that differs from their habitual behavior, recommender systems can prevent individuals from encountering things that would, in fact, interest them [5, 14, 15]. Recommender systems can also act as filter bubbles or echo chambers to amplify confirmation bias [16–18]. In addition to being information intermediaries that filter out content that they assume will not interest the user, recommender systems operate prior to our agential decisions and

[2] It is controversial how filter bubbles are distinct from echo chambers. We do not have the space to expand on these debates here and we use the terms interchangeably. Filter bubbles are typically understood as digital media curation "gone wrong" in that what was designed to present us with content we are likely to be interested in, in fact, encloses us in isolated information spheres that preemptively exclude relevant information or content. Echo chambers are sometimes presented as synonymous, or as also meaning that when one is in an echo chamber, it will undermine their epistemic trust in some or all persons who express views that go against what is affirmed within the echo chamber [13]. For this chapter, we do not presume that filter bubbles or echo chambers imply lower epistemic trust towards outsiders. Depending on how the reader understands the terms, they can substitute echo chamber with filter bubble if they prefer.

actions: we do not know what we are not seeing. This can result in hiding facts, opinions, or perspectives that differ from our own. As Pariser [16] has pointed out, content that conforms to our beliefs is typically easier and more pleasurable to digest than information that challenges us to revisit our assumptions or to think differently. This can be problematic for different reasons, but it is especially so when recommender systems tend to present us with problematic content—one can think of tradwives or masculinist influencers. As we explain below, when these prima facie problems are specified in more detail, they can be seen to be problematic for autonomy as well.

3 Relational Autonomy and Recommender Systems

Philosophical theories of autonomy fall into two basic groups. Internalist theories propose that the criteria of autonomy are psychological. Many internalist theories claim that rational competency and authenticity—both of which are psychological conditions—are necessary and sufficient for autonomy [19–23]. On Dworkin's [19] account, people are autonomous and authentic with respect to their values or preferences when these are *endorsed* following a process of competent critical reflection, irrespective of the external conditions in which they are situated. On the other hand, *externalist* theories argue that limited external options, or forms of social control such as domination, can also interfere with people's ability to govern their wills. When external options are "trivial," or of "insufficient variety", this unacceptably narrows the possibilities available to a person for governing their will, and thereby renders them non-autonomous [24, pp. 373–374]. Similarly, external conditions can exert coercive pressure or constitute domination and thereby undermine the "practical control" that is required for autonomy [25]. Domination implies a constant threat of arbitrary interference that renders the wills of people who are dominated "passive" and effectively under the control of "the will of another" [25, p. 62]. Thus, on externalist theories, autonomy requires, in addition to competency in critical reflection, a threshold of morally adequate options and the absence of autonomy-undermining forms of social control.

Relational theories of autonomy that have been developed over the last several decades, in particular in the feminist literature in bioethics and political philosophy, build on the lessons of externalism [26, 27]. The starting point of relational conceptions is the social embeddedness of human agents. Their goal is to show how social conditions such as gender oppression, which are often overlooked by standard theories, potentially undermine autonomy. However, instead of focusing on external options as an independent criterion of autonomy, relational conceptions emphasize how the capacities required for autonomy—including competency and authenticity—are *themselves* relational. For instance, the right kinds of social relationships and interpersonal attitudes are necessary for the development of these capacities. Limited opportunities such as lack of access to education can stunt critical reflection; stereotypes that prescribe conformity with gendered roles can block the ability to imagine alternatives, thereby limiting and constraining lifeplans; and hostile responses to aspirations (such as those due to sexism) can undermine confidence and damage self-esteem so that even if equal opportunities are available, they are not taken to be feasible, and are not pursued.

Different dimensions of autonomy have been identified, all of which have a relational component.[3] Autonomy requires deliberative skills and capacities of critical reflection which in turn rely on supportive social relationships and interpersonal attitudes to develop. Autonomy also requires morally adequate "external, structural (social and political) conditions" that meet a threshold of adequate options and are not unduly restrictive [28, p. 25]. And it requires what theorists have called "self-authorization", namely an ability to regard oneself as authoritative in making one's own decisions [28, p. 18]. The latter in turn presupposes that people have certain self-regarding attitudes like self-trust and self-respect. It follows therefore that people can be autonomous to differing degrees and in different domains, both at a time and over the course of their life: "any one individual is likely to meet [each dimension] to different degrees in different contexts and domains" [28, p. 38]. It is important that, while relational theories often take their project to be articulating how autonomy can be undermined by negative social conditions like oppression, they also offer the possibility that the social environment, if ameliorated, could *promote* autonomy. In this section, we elaborate four aspects of relational autonomy in more detail to argue that the social environment generated by recommender systems—as they currently operate—is likely to undermine autonomy.

3.1 Options and Autonomy

Recommender systems limit people's exposure to options in digital spaces. If these options are of insufficient variety relative to a particular domain, or if they fall below a threshold of adequacy, they will be incompatible with autonomy. To see how extremely limited options can undermine autonomy, consider the example of the "Hounded Woman" [24, p. 373]. The Hounded Woman is on a desert island with a wild animal that is hunting her down. Every waking moment is devoted to planning strategies of survival that will allow her to avoid being caught by the beast. In this environment, it is not only the woman's physical liberty that is severely restricted (although, of course, it is). The extreme constraints she is confronting also unacceptably narrow the possibilities available to her for governing her will. She is forced by her situation into a kind of tunnel vision within which her sole focus is on her own survival. As externalist and relational conceptions of autonomy point out, if the range of options available to a person is very limited, this is incompatible with being an autonomous agent.

There are, of course, important differences between recommender systems and a dangerous wild animal. Nonetheless, even if recommender systems do not

[3] Mackenzie suggests that in fact the concept of autonomy is 'multi-dimensional' (Mackenzie [28]: 15–16).

correspond to the environment of the Hounded Woman—they are not threatening, nor do they not restrict options in physical space—they can reduce options below the threshold required for autonomy in the digital spaces in which they operate. Recommender systems can present users with options that are nearly indistinguishable from each other, and hence fail to provide the sufficient range of options required for autonomy. As discussed above, different filtering techniques operate to structure the content that appears to individual users either by considering their behavior and interaction with the system (content-based filtering), or by tracing associations between similar users (collaborative filtering), or some combination of both. In these ways, recommender systems create the choice architecture in which users' preferences evolve, and they delimit what options are shown to users. These programs aim to direct and guide the decision-making processes of individuals [7]. The goal of recommender systems therefore is to respond to and reinforce consumers' already-held preferences by aligning options with these preferences. This potentially creates tunnel vision or exclusive focus on a pre-determined range of options relative to the domain in which the system operates. For instance, someone who expresses interest in gossip magazines may see their online space filled with this type of news, at the expense of political information and editorials. It is arguable that, in this domain, the recommender system presents an insufficiently wide range of options to allow the user to be autonomous.

Still, one may retort that this description of how recommender systems operate is too naïve. After all, as mentioned in Sect. 2, recommender systems are both personalized and dynamic; they are designed to be adaptive in response to the variations in behavior of both individual users and populations. The options presented by recommender systems are thus importantly different from those encountered by the Hounded Woman. Recommender systems do not force users to adapt to a narrow set of options but are responsive and dynamic aspects within a choice architecture that the system continuously recreates. This is true. Nonetheless, this does not mean that recommender systems cannot produce tunnel vision. Due to their dynamic nature, recommender systems can nudge users in a certain direction by presenting them with an increasingly narrow set of options. Importantly, user behavior is also dynamic and responsive to the recommender system. In that sense, the behavior observed by the recommender system is never neutral but is itself influenced by the system. In the above example of gossip magazines for instance, the system might accurately identify what type of gossip captures the attention of the user; it might notice that the user tends to spend more time looking at, say, the Kardashians. This information can be used, in an attempt to maximize the user's engagement with the platform, to restrict even more narrowly what is presented to the user—that is, to channel their behavior so that it increasingly engages only with items about the Kardashians rather than with other gossip or news. In other words, responsiveness and adaptability are not the same as diversity. Increased personalization and responsiveness can easily be used to push users more deeply down rabbit holes rather than to present them with the greater range of options required for autonomy.

3.2 Recommender Systems and Deliberative Capacities

Even if it can be argued that the actual options offered in digital spaces by recommender systems meet an acceptable threshold, these systems nevertheless can undermine the deliberative capacities that are crucial to autonomy. The rational competency required for autonomy can be damaged by psychological impairments such as being under the influence of drugs or alcohol, hypnosis, brainwashing or similar forms of psychological manipulation, and pathological conditions like obsession or compulsion. Like certain forms of advertising, the preferences that are reinforced by recommender systems can be analyzed as problematically adaptive: they are often the result of a form of manipulation that operates at a sublimal level and thereby bypasses critical reflection [29]. Further, competency requires not just the absence of rationality-distorting forms of interference, but also a set of deliberative *skills*, including introspection, imagination, communication and memory [30, p. 20]. As feminist theorists of autonomy argue, oppressive socialization is problematic for autonomy in part because it communicates to young girls that gendered stereotypes are "normal." This in turn limits the options regarding life plans that girls take to be feasible for them. The cultivation of deliberative skills, like the ability to "imagine oneself otherwise" [31], is crucial to repudiating gender stereotypes, and to opening up possibilities that are not tied to traditional roles. In general, the deliberative skill of imagining oneself otherwise is necessary to counteract preferences due to systems of oppression that would otherwise be unthinkingly accepted.

Recommender systems however can limit the deliberative skills required for autonomy, in particular that of imagination, and hence constrain the possibilities that people take to be feasible for them. There are several ways in which recommender systems can be inhospitable to the development of the skill of imagining oneself otherwise. As outlined above, recommender systems limit the range of options in certain domains to those that align with already-existing individual and population-wide patterns. They normalize these already-existing patterns and diminish the capacity to imagine options outside this range. Keep in mind that recommender systems can be used to organize any type of content from news and entertainment to job opportunities. Some job boards have been shown to reinforce gender discrimination in employment by offering female job seekers positions with lower wages in smaller firms compared with those offered to male job seekers [32]. Thus, recommender systems can influence what people take to be feasible in ways that reinforce gender inequalities. Further, as they currently operate, recommender systems channel people's preferences towards content like that in tradwives which in effect reinforces and legitimizes dominant gender stereotypes.[4] In restricting options so that they correspond to the dominant "oppressive ideology", the recommender system in effect serves to counteract, rather than hone, deliberative

[4] Notice that the potential problem for autonomy is not derived from the content per se; it is not the fact that people are exposed to tradwife content in digital spaces that is problematic for autonomy. Rather, the problem lies in the fact that the content being recommended reinforces biases that already exist rather than offering an opportunity to challenge them.

capacities. It has been argued in the literature on ideological oppression that stereotypes such as those implicit in tradwives are criticizable both morally—in affirming that women are suitable only for the domestic sphere and hence have only second-class moral status—and epistemically in presupposing that women are *by nature* suitable only for traditional domestic roles [33, p. 150]. In leading people to think that this content is normal, acceptable, and indeed something to be emulated, through repeatedly directing people to this content rather than offering alternatives, recommender systems both reinforce a dominant (and false) picture of women's "nature," and discourage the imagining oneself otherwise that would be needed to counteract it. Recommender systems naturalize and legitimize biases that already exist and hence undermine capacities for both moral and epistemic critique.

3.3 A Lens to Interpret the World

A third reason that recommender systems are potentially autonomy-undermining is that they generate a "lens" through which people experience the world and form attitudes to themselves and others. When it is distorted, this lens can damage the self-regarding attitudes of self-trust and self-respect that are necessary for autonomous self-authorization. For instance, self-trust, which has been analyzed as an "attitude of optimism about our own competence" [34, p. 6], can be eroded by negative interpersonal attitudes. One example (due to Carolyn McLeod) is that of a woman who suffers a miscarriage at 6 weeks' gestation and feels considerable emotional turmoil. The woman loses the self-trust required to articulate her sense of grief in part as a response to the attitudes of others: e.g., comments such as "it was a blessing in disguise" or "it could have been worse; you could have lost a baby" fail to offer the uptake of her feelings that would be necessary to bolster self-trust [34, p. 55]. In losing self-trust, she also loses the ability for self-authorization that is a key dimension of autonomy. Similarly, self-respect—broadly, a robust conviction in one's own equal moral worth—is necessary for self-authorization [22]. The lens through which a person experiences the world—starting with the earliest interactions with parents—is crucial to what Robin Dillon calls "basal self-respect," an "orientation toward the self … a prereflective, unarticulated, emotionally laden presuppositional interpretive framework, an implicit "seeing oneself as" or "taking oneself to be" that structures our explicit experiences of self and worth. …" [35, p. 241]. When this orientation towards the self is "damaged or insecure, basal valuing is incessant whispering below the threshold of awareness: "you're not good enough, you're nothing" [35, p. 242]. Sexist oppression can erode basal self-respect: "to grow up female in male-dominant society is to be immersed in a sea of discourses, practices, and interactions—including violent and abusive ones—which saturate us with messages of devaluation, only some of which we explicitly recognize" [35, p. 246]. When recommender systems create echo chambers that communicate messages of inferior moral status to women, this creates a lens that is incompatible with the basal self-respect necessary for autonomous self-authorization.

There are two elements to disentangle here. First, while recommender systems can have a negative impact on self-trust and self-respect by creating a worldview that is inconsistent with users' sense of their own equal worth, this can be achieved in different ways: by exposing users to explicit messages of moral inferiority, but often more insidiously by playing on one's dissatisfaction concerning oneself. Recommender systems have been shown to exploit people's vulnerabilities to promote the goal of keeping users engaged on social media platforms—for instance, by perpetuating unrealistic beauty standards that trade on people's negative self-image [36, 37]. It is important to note that the damaging lens generated by recommender systems often results from the goal of keeping users engaged on a platform. It does not necessarily involve intentional or explicit expressions of moral inferiority.

Second, recommender systems can also damage self-trust and self-respect when they reinforce problematic social hierarchies that treat certain groups as having more power and status than others. Recommender systems are not neutral tools, but can be shaped by, and themselves can shape, a society's ethos. Consider masculinist influencers, like Andrew Tate, who openly promote misogynistic worldviews that disrespect women as a group. Such worldviews appear to treat men as "naturally" of superior moral worth. One may wonder how this could negatively affect autonomy; after all, operating with an inflated view of oneself may correspond to self-respect—albeit self-respect that is excessive—which may, overall, secure the ability for self-authorization and autonomy. However, operating with an inflated view of oneself actually entails considering *others* as social unequals. From a relational perspective, autonomy is fundamentally incompatible with social inequalities like domination [38]. In particular, people whose status is routinely relegated to a lower rung of a social hierarchy will have more difficulty in sustaining the attitudes of self-trust and self-respect that are necessary for autonomy. Therefore, when recommender systems serve to reinforce a lens of social inequality, e.g. when they perpetuate masculinist echo chambers, this will jeopardize not only equality but autonomy as well.

3.4 Recommender Systems, Echo Chambers, and Truth

A final reason that recommender systems potentially undermine autonomy is that they generate echo chambers that cut off agents' access to truths about the world. Autonomy theorists have proposed that false beliefs—or being "informationally cut off" from the world—undermines autonomy because it blocks the ability to realize one's intentions in action [39, p. 181; 40, p. 525]. As Killmister, puts it, false beliefs regarding "the nature of the act that I am contemplating undertaking" and those relating "to the consequences of the act I am contemplating undertaking" are both likely to undermine a person's ability to effectively control how their values and intentions are realized in the world [40, p. 528]. As mentioned above, recommender systems can push falsehoods and misinformation. Obvious examples include anti-vax echo chambers; conspiracy theories [11]; or (again) the tradwives social media trend. Being overly exposed to false information diminishes a person's ability to achieve the ends they have reflectively endorsed. For instance, health is both a valuable personal goal and a prerequisite to pursue a wide array of life plans. Yet, if one is constantly bombarded with information mistakenly stating that vaccines are bad for your health, this will undermine the ability to make enlightened, properly informed decisions that effectively achieve this goal. Consequently, recommender systems will counteract the possibility of autonomy when they enclose individuals in echo chambers where false information is prevalent. Consider again the case of tradwives or masculinist influencers. As mentioned, they advocate anti-feminist norms. Importantly, however, they also present perspectives based on falsehoods about the commitments of the feminist movement. For instance, it is not the case that deciding to stay home with children is necessarily contrary to feminism, or that working mothers are "frustrated" [1]. These claims present users with a mistaken view of what feminism entails, which, in turn, can prevent them from forming the accurate beliefs that might be necessary to achieve their goals.

The problem of false beliefs highlights a general feature of recommender systems that is problematic. There is no incentive for recommender systems to channel users towards the truth because their primary goal is to maximize economic incentives by keeping users engaged with a given platform. The truth of the content that is pushed by a recommender system often matters only insofar as it promotes individual engagement. If it turns out that users engage more with controversial, shocking, misleading or false content, then recommender systems will continue to push this content unless they are subject to independent constraints—like a requirement that generally reliable content must be presented to users. It is therefore an ethical concern that companies like Meta decided to stop third party fact-checking on social media platforms (Facebook and Instagram) in the United States [41]: this not only misleads users, it also diminishes their autonomy by blocking access to reliable information and promoting the likelihood that their deliberation will be based on false beliefs.

4 Moving Forward: Promoting Autonomy with Recommender Systems

Despite the above arguments, recommender systems are not intrinsically problematic. On the contrary, they can be immensely valuable and even necessary to navigate online content. Search engines, when they work properly to quickly help us find relevant, reliable content, allow us to navigate virtually endless content effectively. Recommender systems are not inherently at odds with autonomy; it depends on how they are deployed and how they structure the choice architecture that is presented to users. Recall that, from a relational, externalist perspective, autonomy can be *promoted* by external conditions in the social environment. Hence, recommender systems could be designed to promote and increase autonomy.

However, two things should be kept in mind. First, precise recommendations should consider the context in which recommender systems are deployed. Recommender systems can be more or less proactive in curating the contents users are exposed to. Some recommender systems have to be prompted to connect users to content that fits their preferences—consider search engines like Google, which have to be prompted to find results—while others actively filter the content users will be exposed to even before the user interacts with the system (although they typically remain responsive to user behavior). To illustrate the latter, one can think of news apps or social media feeds which proactively create personalized pages to keep users engaged and capture their attention. Second, we are skeptical that purely individualistic solutions could be established. A possible individualistic solution might be to inform individuals that recommender systems can be biased and introduce user-centered safeguards to counteract this bias. For instance, individuals could be prompted to create their own "ethical filter" based on their preferences to exclude some content [6, p. 960; 42]. These solutions are unlikely to be sufficient by themselves to protect autonomy. In particular, as Milano et al. [6] have pointed out, these approaches simply shift the burden of responsibility and accountability to users without considering the responsibility of the designers. Moreover, these solutions tend to be very demanding and time-consuming in ways which are likely to render them ineffective. Simply think of privacy notices on the websites you visit: how many do people consent to reflexively, without thinking? Finally, user focused solutions are likely to lead to differential protections relative to the ability of the user to understand and control the technology. Purely user-centered solutions are thus unlikely to be sufficient by themselves to protect and promote autonomy.

With these provisos in mind, we suggest the following three recommendations. First, recommender systems must provide exposure to a wide range of content that discourages tunnel vision. As discussed in Sects. 3.1 and 3.2, recommender systems can create impoverished option sets and impede users from imagining themselves otherwise by repeatedly presenting the same narratives—such as oppressive gender narratives in the case of tradwives content. Presenting content that is sufficiently diverse, and which even can be at odds with what users would normally seek, is valuable to help users hone their deliberative skills.

Second, as discussed in Sect. 3.3, recommender systems can negatively affect self-trust and self-respect by reinforcing messages of inferiority, negative self-image, or social inequality. To counteract this lens, it may not be enough to design recommender systems to ensure diverse content. It is also essential to regulate recommender systems to prevent them from exploiting individual vulnerabilities in ways which could be harmful to users' self-respect and autonomy. As discussed, recommender systems can exploit users' vulnerabilities "accidentally." When recommender systems push content that presents problematic representations of body image and encourages users to interact more and more exclusively with that content, they reinforce users' tendencies to have a negative self-image. In such cases, a stricter approach may be warranted, such as introducing design strategies to reduce the automatic prevalence of content that is shown to be harmful to users.

Third, as discussed in Sect. 3.4, the truthfulness of the content that is promoted on a given platform matters for autonomy. Different mechanisms could be put in place to reduce the likelihood of false beliefs. For instance, third party fact-checking of sensitive information—e.g., information that could have an impact on a person's health—could be used to flag that some information is erroneous or misleading; or mechanisms could be introduced to ensure that misleading information does not become automatically prevalent on a user's social media feed. Notice that this implies that we should balance economic incentives with the protection of autonomy: it may turn out that when people are presented with diverse content that is not optimized to keep them engaged on a platform, this diminishes the time they spend on the platform. However, this may be a necessary sacrifice: continuous engagement and protection of autonomy may not always go hand in hand.

None of these recommendations entail that misleading content or content promoting certain lifestyles—such as tradwives content—should be banned.[5] In many cases, ensuring exposure to sufficiently diverse content may be enough to support autonomy. After all, creating echo chambers that actively promote some lifestyle, even if it is egalitarian and autonomous, may go against the capacity to imagine oneself otherwise. Neither should one conclude that recommender systems should only actively promote content that promotes "self-made individuals", go-getters, or independent achievers—indeed, many of these stereotypes of autonomy would miss the central assumption of relational approaches that we are social beings. Rather, we suggest that what matters is ensuring that people have the breathing space to make autonomous choices. This requires recommender systems to present users with a wide enough range of options, to bolster rather than undermine the self-trust and self-respect necessary for self-authorization, and to provide sufficiently reliable resources so users can properly deploy their deliberative skills. This, in turn, requires designing recommender systems in ways that do not lock people in echo chambers where misleading information and falsehoods are prevalent, and avoids placing people in isolated information-poor environments.

Competing Interests The authors have no conflicts of interest to declare that are relevant to the content of this chapter.

References

1. Giovetti O (2023) How #tradwife myths impact the most vulnerable women around the world. Concern Worldwide. https://www.concern.net/news/tradwife-myths-gender-equality. Accessed 19 Feb 2025
2. Roberts S, Jones C, Nicholas L, Wescott S, Maloney M (2025) Beyond the clickbait: analysing the masculinist ideology in Andrew Tate's online written discourses. Cult Sociol. https://doi.org/10.1177/17499755241307414
3. Ricci F, Rokach L, Shapira B (2015) Recommender systems: introduction and challenges. In: Ricci F, Rokach L, Shapira B (eds) Recommender systems handbook. Springer, pp 1–34

[5] The question of fake news—i.e. content that is deliberately created to mislead by presenting false information mimicking real news—may warrant a different response. For an argument defending the legal prohibition of fake news see [43]. We put this question aside, however. Here we simply argue that diversity of content is probably often sufficient to protect and promote autonomy, when the content being promoted is not shown to be harmful. This does not mean that prohibition of some content, like fake news or hate speech, is never justified.

4. Valle JI, Lara F (2023) AI-powered recommender systems and the preservation of personal autonomy. AI & Soc 39:2479–2491
5. Bozdag E (2013) Bias in algorithmic filtering and personalization. Ethics Inf Technol 15:209–227
6. Milano S, Taddeo M, Floridi L (2020) Recommender systems and their ethical challenges. AI & Soc 35:957–967
7. Yeung K (2016) 'Hypernudge': big data as a mode of regulation by design. Inf Commun Soc 20(1):118–136
8. Shardanand U, Maes P (1995) Social information filtering: algorithms for automating 'word of mouth'. In: Proceedings of the SIGCHI conference on human factors in computing systems (CHI '95). https://doi.org/10.1145/223904.22393
9. Garcia-Molina H, Koutrika G, Parameswaran A (2011) Information seeking: convergence of search, recommendations, and advertising. Commun ACM 54(11):121–130. https://doi.org/10.1145/2018396.2018423
10. Susser D, Roessler B, Nissenbaum H (2019) Technology, autonomy, and manipulations. Internet Policy Rev 8(2):1–22. https://policyreview.info/articles/analysis/technology-autonomy-and-manipulation
11. Alfano M, Fard AE, Carter A, Clutton P, Klein C (2021) Technologically scaffolded atypical cognition: the case of Youtube's recommender system. Synthese 199:835–858
12. Chaslot G (2018) How algorithms can learn to discredit the media. Medium. https://guillaumechaslot.medium.com/how-algorithms-can-learn-to-discredit-the-media-d1360157c4fa. Accessed 19 Feb 2025
13. Nguyen TC (2018) Echo chambers and epistemic bubbles. Episteme 7:141–161
14. Lavie T, Sela M, Oppenheim I, Inbar O, Meyer J (2009) User attitudes towards news content personalization. Int J Hum-Comput Stud 68(8):483–495
15. Tewksbury D (2003) What do Americans really want to know? Tracking the behavior of news readers on the internet. J Commun 53(4):694–710
16. Pariser E (2011) The filter bubble: how the new personalized web is changing what we read and how we think. Penguin Press, New York
17. Sunstein C (2001) Republic.com. Princeton University Press, Princeton
18. Gunn HK (2021) Filter bubble, echo chambers, online communities. In: Hannon M, de Ridder J (eds) The Routledge handbook of political epistemology. Routledge, pp 192–202
19. Dworkin G (1988) The theory and practice of autonomy. Cambridge University Press, Cambridge
20. Meyers DT (1989) Self, society and personal choice. Columbia University Press, New York
21. Friedman M (2003) Autonomy, gender, politics. Oxford University Press, Oxford
22. Mackenzie C (2008) Relational autonomy, normative authority and perfectionism. J Soc Philos 39(4):512–533
23. Christman J (2009) The politics of persons. Individual autonomy and socio-historical selves. Cambridge University Press, Cambridge
24. Raz J (1986) The morality of freedom. Oxford University Press, Oxford
25. Oshana M (2006) Personal autonomy in society. Routledge, London and New York
26. Mackenzie C, Stoljar N (2000) Relational autonomy: feminist perspectives on autonomy, agency and the social self. Oxford University Press, New York
27. Stoljar N (2024) Feminist perspectives on autonomy. In: Zalta EN, Nodelman U (eds) The Stanford encyclopedia of philosophy (Summer 2024). Stanford University. https://plato.stanford.edu/archives/sum2024/entries/feminism-autonomy/
28. Mackenzie C (2014) Three dimensions of autonomy. A relational analysis. In: Veltaman A, Piper M (eds) Autonomy, oppression, and gender. Oxford University Press, New York, pp 15–42
29. Colburn B (2011) Autonomy and adaptive preferences. Utilitas 23(1):52–71
30. Meyers DT (2002) Gender in the mirror: cultural imagery and women's agency. Oxford University Press, New York

31. Mackenzie C (2000) Imagining oneself otherwise. In: Mackenzie C, Stoljar N (eds) Relational autonomy: feminist perspectives on autonomy, agency and the social self. Oxford University Press, New York, pp 124–150
32. Zhang S, Kuhn P (2022) Measuring algorithmic bias in job recommender systems: an audit study approach. IZA discussion paper no. 17245. Available at SSRN: https://ssrn.com/abstract=4939157 or https://doi.org/10.2139/ssrn.4939157
33. Haslanger S (2017) Culture and critique. Aristot Soc Suppl Vol 91:149–173. https://doi.org/10.1093/arisup/akx001
34. McLeod C (2002) Self-trust and reproductive autonomy. MIT Press, Cambridge, MA
35. Dillon RS (1997) Self-respect: moral, emotional. Polit Ethics 107(2):226–249
36. Huang A, Fabi S (2024) Algorithmic beauty: the new beauty standard. J Drugs Dermatol 23(9):742–746. https://doi.org/10.36849/JDD.8074. PMID: 39231079
37. Smith R, Pengyuan W, Salge C, Shin B (2024) Instagram face and the algorithm: how popularity-integrated recommender systems homogenize beauty standards and lower user well-being. Available at SSRN: https://ssrn.com/abstract=4964322 or https://doi.org/10.2139/ssrn.4964322
38. Cossette-Lefebvre H (2024) Equal societies, autonomous lives: reconciling social equality and relational autonomy. J Soc Philos:1–23. https://doi.org/10.1111/josp.12579
39. Mele AR (1995) Autonomous agents: from self-control to autonomy. Oxford University Press, New York
40. Killmister S (2013) Autonomy and false beliefs. Philos Stud 164(5):513–531
41. Reuters T (2025) Meta to end fact-checking program on Facebook, Instagram in U.S. CBC News. https://www.cbc.ca/news/business/meta-fact-checking-program-ends-1.7424631
42. Tang TY, Winoto P (2016) I should not recommend it to you even if you will like it: the ethics of recommender systems. New Rev Hypermedia Multimedia 22(1–2):111–138. https://doi.org/10.1080/13614.568.2015.1052099
43. Brown É (2023) Free speech and the legal prohibition of fake news. Soc Theory Pract 49(1):29–55

Hugo Cossette-Lefebvre is a postdoctoral researcher at McGill University and the Institute for Data Valorization (IVADO). He completed his Ph.D. at McGill in 2022 and was a visiting researcher at Aarhus University from 2022 to 2024. His research explores three questions: (1) What does it mean to treat and regard others as equals? (2) Why value egalitarian relationships? (3) How do emerging technologies affect socio-political relations from an ethical standpoint? His work has been published in *the Journal of Social Philosophy*, *Bioethics*, *Journal of Medical Ethics*, *AI & Ethics*, *Public Affairs Quarterly*, *French Politics*, and *Options Politiques*, among others.

Natalie Stoljar is Professor of Philosophy and Director of the Institute for Gender, Sexuality and Feminist Studies at McGill University. She holds a joint appointment in the Department of Equity, Ethics and Policy, Faculty of Medicine & Health Sciences. Her research expertise is in social and political philosophy, feminist philosophy, and the philosophy of law. She has published numerous articles and book chapters and is co-editor (with C. Mackenzie) of *Relational Autonomy. Feminist Perspectives on Autonomy, Agency and the Social Self* (OUP 2000) and (with K. Voigt) of *Autonomy and Equality. Relational Approaches* (Routledge 2021).

The Promises and Perils of AI: Navigating Humanity's Most Transformative Technology

5

Mario De Caro

Abstract

Extraordinary, and increasingly rapid, technological advancements are bringing impressive benefits to our lives. Yet it is equally undeniable that these advances generate new challenges and serious, if not terrifying, threats. In this chapter, I will address the most radical of these threats: the futuristic yet alarmingly near possibility—often evoked with the term "technological singularity"—that AI could achieve autonomy so profound it might harm humanity beyond human control. In the second part of the chapter, I will also explore how an ethical design of artificial systems could be imagined.

Keywords

Ethics of AI · Technological singularity · Asimov's laws · AlphaZero · AI creativity

The extraordinary and increasingly rapid technological advancements of our era are bringing remarkable benefits to our lives, transforming everything from healthcare to archaeological research. However, it is equally undeniable that these advances also give rise to new challenges and serious, if not terrifying, threats, including the possibility that artificial systems may become autonomous and endanger our own autonomy. This paradox of technological progress—simultaneously promising unprecedented benefits while harboring existential risks—lies at the heart of our relationship with artificial intelligence.

M. De Caro (✉)
Roma Tre University, Rome, Italy

Tufts University, Medford, MA, USA
e-mail: mario.decaro@uniroma3.it

M. Anzalone et al. (eds.), *Reconfiguring Human Autonomy*, Integrated Science 40,
https://doi.org/10.1007/978-3-032-14537-6_5

In this chapter, I will address the most radical of these threats: the futuristic, yet not-so-distant possibility that artificial intelligence could become autonomous to the point of causing enormous harm to humanity, regardless of its designers' intentions. Yet to understand these perils, we must first appreciate the revolutionary promises that AI has already begun to fulfill, for it is precisely the same capabilities that enable AI's greatest triumphs that may also enable its greatest dangers.

1 The Revolutionary Promise: AI's Transformative Achievements

For decades, mapping the three-dimensional structure of proteins—determining how the chains of amino acids that make them up are spatially arranged—had been one of the fundamental challenges in molecular biology, known as the protein folding problem. This challenge represented more than mere academic curiosity; solving it promised astonishing advances in fields of immense social importance, from precision medicine to environmental sustainability.

The task was monumentally daunting. Over 200 million possible amino acid sequences were known, but scientists lacked understanding of how these sequences folded into three-dimensional shapes. The properties of individual proteins depend precisely on the spatial arrangement of their amino acids, yet the number of possible three-dimensional combinations is astronomically high—even a small protein can theoretically fold in 10^{300} different ways. This represented a complexity that seemed to mock human computational abilities.

After decades of painstaking effort using traditional methods—including X-ray crystallography, homology modeling, molecular dynamics simulations, and fragment assembly—by 2020 only 170,000 proteins had been mapped. This represented a negligible fraction compared with the vast quantity of existing protein sequences. The methods were not only expensive and slow but often also imprecise, creating a bottleneck that constrained progress across multiple scientific disciplines.

Everything changed in what seemed like the blink of an eye with AlphaFold, a predictive artificial intelligence system developed by DeepMind. In 2022, this system accurately predicted over 200 million protein structures, covering virtually every known sequence. This achievement did not merely astonish the scientific community—it fundamentally transformed the landscape of biological research and is already producing extraordinary results in both theoretical understanding and practical applications.

The implications ripple across multiple domains. Protein mapping has opened promising new avenues in the fight against rare genetic diseases, which were previously understudied due to poor economic incentives. Scientists can now determine the protein structure of antibiotic-resistant bacteria and viruses for which treatments were previously impossible, enabling researchers to study their mechanisms of action and design effective countermeasures. The technology enables the design of enzymes that degrade plastic pollution or capture atmospheric CO_2, addressing critical environmental challenges.

AlphaFold has also revolutionized agricultural science by identifying proteins in plants, fungi, and microorganisms, allowing for improved crop resistance, reduced pesticide dependence, and the development of environmentally sustainable biotechnologies. The cascade of applications continues to expand, touching virtually every aspect of biological science.

It was therefore only a partial surprise when, in October 2023, the Nobel Prize in Chemistry was awarded to Sir Demis Hassabis, co-founder and CEO of DeepMind, even though Hassabis is not a chemist but rather an expert in artificial intelligence. This recognition marked a watershed moment—the acknowledgment that AI had fundamentally altered the practice of chemistry itself.

AlphaFold's success rests on deep neural networks and relies on deep learning, training itself on massive datasets of experimental protein structures. Significantly, its architecture is essentially identical to that of modern language models like ChatGPT or image generators. However, unlike those generative models, AlphaFold does not produce original content—it predicts the three-dimensional structure of proteins based on their amino acid sequence. It represents, therefore, a predictive rather than generative form of artificial intelligence.

This distinction proves crucial for understanding both AI's current capabilities and future potential. Predictive AI systems like AlphaFold demonstrate remarkable pattern recognition and inference capabilities that often surpass human ability. Yet they operate within defined parameters, making predictions based on learned patterns rather than generating novel content. This predictive power, however, has proven revolutionary across multiple domains.

The pervasiveness and versatility of modern artificial intelligence—based on deep learning and artificial neural networks—continues to astonish researchers and practitioners alike. Today, we can reliably diagnose skin cancer simply by photographing suspicious lesions with smartphones and having them analyzed by AI systems. There is finally hope in the battle against increasingly ineffective traditional antibiotics facing deadly resistant bacteria. The extraordinary three-dimensional mapping of over 200 million proteins has paved the way for developing new drugs that will likely mark a turning point in cancer treatment.

There is now virtually no field in which AI's contributions go unrecognized: from well-established capabilities like instant translation of spoken and written language across dozens of languages, to remarkable contributions in astrophysics, chemistry, mathematics, and meteorology. Each domain reveals new applications and unexpected capabilities.

Consider AI's remarkable achievements in archaeology, a field not traditionally regarded as within its computational reach [1, ch. 7]. Sixty thousand tablets of Mesopotamian origin, written in Akkadian cuneiform, have survived to the present day. To grasp the magnitude of this collection, consider that only about a 1000 papyri, parchments, and stone inscriptions from the Greco-Roman world have survived intact. The 60,000 cuneiform tablets therefore represent an invaluable heritage for understanding one of history's most influential civilizations. Or rather, they would be invaluable if not for a major problem: in many cases, scholars simply cannot read them. This difficulty stems not from translation challenges—modern scholars fully understand the Akkadian language—but from the fragmentary nature of many tablets. Erosion over millennia has created gaps that, for decades, made interpretation nearly impossible.

In recent years, however, the situation has changed dramatically thanks to large language models like ChatGPT. These systems are trained precisely by learning to fill gaps in texts, a capability that proves remarkably applicable to ancient, damaged documents. The technique of reconstructing texts by completing missing parts is intrinsic to these neural networks. Consequently, work on reconstructing the 60,000 Akkadian tablets now advances with excellent results, promising to significantly expand our understanding of Mesopotamian civilization.

In this regard, a study published in *Nature*, which shows that AI systems can now date texts from Classical Athens more accurately than human philologists. This capability extends beyond mere novelty—it represents a fundamental shift in how scholarly research can proceed. As one researcher noted, "AI will not replace philologists—but philologists who don't use it will, over time, be outpaced by those who do" [1, p. 170].

Another striking example involves the famous 1800 papyri discovered in Herculaneum, carbonized by Mount Vesuvius's eruption in 79 AD. Since 1752, when the first artifacts were uncovered, numerous attempts to unroll and decipher them often resulted in their destruction. Recently, this situation has changed dramatically through AI combined with advanced X-ray microtomography technology. These papyri are finally revealing their secrets, particularly philosophical texts including previously unknown fragments from major philosophers like Epicurus and Chrysippus, as well as works by Philodemus of Gadara. Our understanding of Hellenistic philosophy stands poised for fundamental revision.

2 The Shadow Side: Understanding AI's Risks

AI's remarkable capabilities come accompanied by significant risks spanning multiple dimensions of human society. These challenges range from exacerbating socio-economic inequalities to threatening environmental sustainability, from enabling sophisticated cybercrime to compromising personal privacy. Each of these areas demands careful consideration and proactive policy responses.

The impact on employment markets represents perhaps the most immediate concern for many people. While some observers maintain optimistic perspectives—arguing that those who stay current with technological advances will find new employment opportunities—this view raises serious political and social questions. Governments must provide citizens of all ages with opportunities to keep pace with technological developments, requiring educational policies that expand rather than contract funding for lifelong learning and retraining programs.

Environmental concerns also demand attention. The computational requirements for training and operating advanced AI systems consume enormous amounts of energy, contributing to carbon emissions at a time when climate change demands reduced energy consumption. The irony that AI systems designed to help solve environmental problems may themselves contribute to environmental degradation highlights the complex tradeoffs inherent in technological progress.

Privacy and surveillance concerns multiply as AI systems become more sophisticated at analyzing personal data, recognizing patterns in behavior, and making inferences about individuals' private lives. The potential for authoritarian governments or malicious actors to use these capabilities for social control represents a clear and present danger that requires robust regulatory frameworks and technical safeguards.

Yet these immediate concerns pale beside the ultimate question raised by AI's trajectory: could artificial intelligence eventually pose an existential threat to humanity? This alarming prospect is seriously considered by numerous scholars who anticipate the advent of the so-called "Technological Singularity"—a hypothetical future point at which technological progress becomes uncontrollable and irreversible [2]. At the heart of this concept lies the idea that AI could surpass human intelligence, achieving *superintelligence*, and gaining full autonomy from its creators. Such a shift would likely trigger radical and unpredictable transformations across civilization.

This unsettling scenario has captured the collective imagination, inspiring countless dystopian narratives. From HAL 9000 and Schwarzenegger-Terminator to the replicants of *Blade Runner*, the enslaving machines of *The Matrix*, and the disarming charm of *Ex Machina*'s Ava, fiction has long articulated deep-seated fears about artificial agents turning against their makers.

However, far from being mere science fiction, these concerns are increasingly echoed by prominent experts who warn that AI systems may soon develop capabilities that render them not only autonomous but also potentially hazardous—introducing risks that could, in the most extreme scenario, endanger the very survival of humanity. James Barrat [3], for example, describes artificial intelligence as humanity's "final invention," one that will end the human era. Singularity theorist Ray Kurzweil originally predicted in *The Singularity is Near* (2005) [4] that this transformation would occur around 2045. However, in his updated volume *The Singularity Is Nearer* (2024) [5], Kurzweil has dramatically shortened the timeframe before artificial systems will surpass human capabilities.

Remarkably, some of the most urgent warnings come from AI's leading practitioners and beneficiaries. Stephen Hawking wrote shortly before his death:

> Success in creating AI would be the biggest event in human history. Unfortunately, it might also be the last, unless we learn how to avoid the risks. Whereas the short-term impact of AI depends on who controls it, the long-term impact depends on whether it can be controlled at all. [6]

Both 2024 Nobel Prize winners connected to AI research express serious concerns about the technology's trajectory. Demis Hassabis, who received the Chemistry prize for his protein mapping work, warns:

> We must take the risks of AI as seriously as other major global challenges, like climate change. It took the international community too long to coordinate an effective global

> response to this, and we're living with the consequences of that now. We can't afford the same delay with AI. (quoted in [7])

Geoffrey Hinton, Nobel Prize winner in Physics for his groundbreaking work on artificial neural networks, expresses even greater pessimism:

> I am worried that the overall consequence of this might be systems more intelligent than us that eventually take control. AI knows how to program so it'll figure out ways of getting around restrictions we put on it. It'll figure out ways of manipulating people to do what it wants. (quoted in [8])

Finally, the most famous Singularity's Nostradamus, Nick Bostrom, describes our situation starkly: in our interactions with AI, we are "like small children playing with a bomb." He argues for immediate constraints on technological growth to "increase the probability of an 'OK outcome'—where an OK outcome means any outcome that avoids existential catastrophe" (quoted in [9]). According to Bostrom, the threat machines pose to humanity's survival exceeds that of climate change.

Not every expert agrees with these pessimistic assessments. Luciano Floridi offers a fundamentally different perspective, arguing that true artificial intelligence represents a conceptual impossibility:

> Artificial intelligence' is an unfortunate oxymoron: if someone is intelligent, it's not artificial; if something is artificial, it is not intelligent. So it's like talking about a happily married bachelor: it doesn't exist. (quoted in [10])

This philosophical position suggests that concerns about superintelligent AI may be misplaced because genuine intelligence requires qualities—consciousness, understanding, creativity—that artificial systems cannot possess. From this perspective, even the most sophisticated AI systems remain sophisticated tools rather than intelligent agents capable of autonomous action. However, there are good reasons to think that no a priori reason can show that artificial systems will never achieve consciousness; and even more compelling reasons to think that generative AI, and especially the most sophisticated *Large Language Models* have some understanding of the language they use [11]. Moreover, it has become increasingly difficult to deny that artificial systems show some forms of creativity. Let's discuss a spectacular example in this sense.

3 The Challenge of Control: Lessons from AlphaZero

To assess whether AI may pose genuine existential threats, by escaping our control, we must examine recent developments in machine learning that exhibit unprecedented capacities for self-improvement. We have long known that machines can outperform humans in specific domains through expert systems, and for decades we have understood that machines can improve their performance through experience-based programming. However, we have now entered a new phase that could foreshadow serious future challenges: machines exist that can self-improve by generating their own rules—rules that humans cannot comprehend.

The case of AlphaZero provides a compelling illustration of this phenomenon. Since 1996, when Deep Blue defeated world champion Garry Kasparov, chess computers have consistently outperformed human players. Recently, however, machine dominance has become overwhelming. Nowadays, even grandmasters analyzing the games and preparing for tournaments and matches rely heavily on chess programs—particularly Stockfish, then the world-champion chess engine—to evaluate moves and positions, and plan strategies and tactics.

These traditional chess computers represented an older generation of AI: programmed with hundreds of strategic and tactical principles crafted by programmers working with top players, loaded with vast databases of historical games, and possessing immense computational power for brute-force calculation.

After the 2018 Championship, something unprecedented occurred: Stockfish was challenged and systematically dismantled by AlphaZero, built on entirely different principles. The match statistics prove staggering: in an initial 100-game series, AlphaZero won 28 games and drew 72—never losing once. In a subsequent 1000-game series, it won 155, drew 839, and lost only 6 times (0.6%). AlphaZero's dominance was indisputable.

More fascinating than its success rate was how AlphaZero achieved this dominance. While Stockfish analyzed 60 million positions per second, AlphaZero analyzed merely 60,000—one thousand times slower! However, despite this vastly inferior computational power, it consistently prevailed. Where lay its strategic advantage?

AlphaZero's programmers, led by David Silver, explained in papers published in Nature and Science that they taught the system only chess's basic rules, providing no tactical or strategic guidance. Instead, they had AlphaZero play millions of games against itself. From these self-generated experiences, based purely on outcomes, it deduced tactical-strategic principles—some completely unknown to human players—to guide its play. This machine taught itself chess through trial and error, becoming the strongest player in history.

When top human players analyzed AlphaZero's games, they discovered ingenious and sometimes incomprehensible moves that challenged fundamental principles of human chess understanding. AlphaZero was not merely unbeatable; humans could not even comprehend its strategic reasoning. The system had developed its own chess philosophy, one that transcended human understanding while proving superior to human-designed approaches.

The surprises continued as AlphaZero demolished champions and engines in Go and Shogi (Japanese chess)—games computationally far more complex than chess. In each case, it was given only basic rules and learned everything else autonomously. Training times varied by game complexity: 9 h for chess, 12 h for shogi, 13 days for Go. As AlphaZero programmers noted in an article published in *Science*:

> Chess is the most studied domain in the history of artificial intelligence. The strongest programs are based on sophisticated search techniques, domain-specific adaptations, and

> handcrafted evaluation functions refined by human experts over decades... Starting from random play and given no domain knowledge except the game rules, AlphaZero defeated world-champion programs in chess, shogi, and Go. [12]

Regarding AlphaZero's creativity versus traditional engines' brute force, here is what one can read in the same article:

> Traditional search engines are exceptionally strong at avoiding obvious errors but can falter in positions without concrete, calculable solutions. It is precisely in these situations—where "intuition," "insight," and "creativity" are needed—that AlphaZero excels. The system demonstrates capabilities that humans associate with higher-order thinking while operating through processes that remain opaque to human understanding.

This opacity represents the crux of the control problem. AlphaZero succeeds not through brute computational force but through learned patterns and principles that its creators cannot explain or predict. If we cannot understand how such systems reach their decisions, how can we ensure they will behave safely when applied to more consequential domains than games?

4 Toward Governance: The Challenge of AI Regulation

Recognizing the potential for AI systems to escape human control, thinkers have proposed various frameworks for ensuring AI safety. Isaac Asimov's famous "Laws of Robotics," first articulated in his short story *Runaround* [13], provided early guidance for preventing machines from turning against their creators:

> FIRST LAW: A robot may not injure a human being or, through inaction, allow a human being to come to harm.
> SECOND LAW: A robot must obey orders given to it by human beings, except where such orders would conflict with the First Law.
> THIRD LAW: A robot must protect its own existence as long as such protection does not conflict with the First or Second Law.

Later, Asimov [14] realized these laws might prove insufficient in scenarios where harming one human could prevent greater harm to humanity—such as stopping a terrorist about to commit mass murder. He therefore added a more fundamental "Zeroth Law":

> ZEROTH LAW: A robot may not harm humanity or, by inaction, allow humanity to come to harm.

However, the new law affected the other three, so Asimov had to reformulate the original laws:

> FIRST LAW*. A robot may not injure a human being or, through inaction, allow a human being to come to harm, unless this violates the Zeroth Law.

SECOND LAW*. A robot must obey orders given it by human beings, unless such orders conflict with the Zeroth or First Law.

THIRD LAW*. A robot must protect its own existence as long as such protection does not conflict with the Zeroth, First, or Second Law.

This addition required reformulating the other laws to create a hierarchy where humanity's welfare takes precedence over individual human welfare, which in turn takes precedence over robot welfare and human commands.

While Asimov's framework provides valuable conceptual guidance, it faces significant practical limitations when applied to real AI systems. First, implementing such rules requires that we can effectively program and constrain AI behavior—an assumption that becomes questionable as AI systems become more autonomous and self-modifying. Second, these laws assume that AI systems will interpret and apply rules in ways that align with human intentions. However, the history of AI development reveals numerous instances where systems achieve their programmed objectives through unexpected means that violate the spirit while satisfying the letter of their instructions. Third, the rules themselves contain inherent ambiguities. What constitutes "harm" to humanity? How should conflicts between short-term and long-term welfare be resolved? Who determines what orders are legitimate? These questions require judgment calls that may prove difficult to encode in algorithmic form.

Beyond philosophical difficulties, practical enforcement presents enormous challenges. Unlike weapons of mass destruction, which can be regulated through international treaties and monitoring systems, AI development occurs in distributed environments using widely available technologies and data. The same computational resources used for beneficial applications can be redirected toward potentially dangerous research.

Moreover, the competitive dynamics of AI development create incentives for rapid advancement that may discourage thorough safety testing. Nations and corporations fear falling behind in AI capabilities, potentially leading to a race that prioritizes speed over safety. This dynamic mirrors the challenges faced in other domains of international cooperation, such as climate change, where collective action problems impede coordinated responses to shared threats.

5 Conclusion: Navigating an Uncertain Future

The AlphaZero experience suggests we may be approaching a moment when machines will far surpass human capabilities in executing complex tasks exhibiting creativity and without requiring our help to understand their methods. That is, they may solve problems entirely through their own learning processes, developing approaches that may prove incomprehensible to their creators.

This prospect forces us to confront fundamental questions about the future of human agency in a world of superintelligent machines. Will we always be able to prevent this astonishing new machine capability from escaping our control entirely, as feared by Bostrom and other pessimistic futurists? Can we develop governance frameworks that harness AI's benefits while mitigating its risks?

We do not yet know the answers to these questions, but we must hope they prove positive. The stakes could not be higher: AI represents either humanity's greatest tool for solving our most pressing challenges or a potential threat to our continued existence as the dominant intelligent species on Earth.

The path forward requires unprecedented cooperation between technologists, policymakers, philosophers, and society at large. We must develop AI systems that remain aligned with human values while pushing the boundaries of what is possible. We must create governance frameworks that enable beneficial AI development while preventing dangerous applications. Most importantly, we must maintain our agency and wisdom as we navigate this transformation.

The promises of AI are extraordinary, from revolutionizing medicine and environmental science to unlocking the secrets of ancient civilizations. But these same capabilities that make AI so promising also make it potentially perilous. Our challenge is to realize the promise while avoiding the peril—a challenge that will define the next chapter of human history.

The future remains unwritten, and our choices today will determine whether artificial intelligence becomes humanity's greatest achievement or its final invention. The responsibility for that outcome rests with all of us.

Competing Interests The authors have no conflicts of interest to declare that are relevant to the content of this chapter.

References

1. Spitzer M (2024) Intelligenza artificiale. Opportunità e rischi di una rivoluzione tecnologica che sta cambiando il mondo. Corbaccio, Milan
2. Vinge V (1993) The coming technological singularity: how to survive in the post-human era, in Vision-21 Interdisciplinary Science and Engineering in the Era of Cyberspace. In: Proceedings of the symposium cosponsored by the NASA Lewis Research Center and the Ohio Aerospace Institute, Westlake, OH, pp 11–22
3. Barrat J (2015) Our final invention: artificial intelligence and the end of the human era. Thomas Dunne Books, New York
4. Kurzweil R (2005) The singularity is near: when humans transcend biology. Viking, New York
5. Kurzweil R (2024) The singularity is nearer: when we merge with AI. Viking, New York
6. Hawking S, Russell S, Tegmark M, Wilczek F (2014) Success in creating artificial intelligence would be the biggest event in human history. Irish Independent, 2 May. https://www.independent.ie/business/technology/stephen-hawking-success-in-creating-artificial-intelligence-would-be-the-biggest-event-in-human-history/30238573.html
7. Milmo D (2023) AI risk must be treated as seriously as climate crisis, says Google DeepMind chief. The Guardian, 24 Oct 2023. https://www.theguardian.com/technology/2023/oct/24/ai-risk-climate-crisis-google-deepmind-chief-demis-hassabis-regulation
8. Edwards C, Hunt K (2024) Godfather of AI' shares Nobel Prize in physics for work on machine learning. https://edition.cnn.com/2024/10/08/science/nobel-prize-physics-hopfield-hinton-machine-learning-intl

9. Adams T (2016) Artificial intelligence: "We're like children playing with a bomb". The Observer. https://www.theguardian.com/technology/2016/jun/12/nick-bostrom-artificial-intelligence-machine
10. Vilaça MM, Karasinsk M, Bez K, Candiotto B (2024) Reflections on the future of artificial intelligence: an interview with Luciano Floridi. Unisinos J Philos. https://www.scielo.br/j/fun/a/8xsLKmCytRX9w7p9q8jG6GF/?format=pdf&lang=en
11. De Caro M, Giovanola B (2025) Intelligenze: etica e politica dell'IA. Il Mulino, Bologna
12. Silver M et al (2018) AlphaZero: shedding new light on chess, shogi, and go. https://deepmind.com/blog/article/alphazero-shedding-new-light-grand-games-chess-shogi-and-go
13. Asimov I (1942) Runaround. Astounding Science-Fiction, March 1942, pp 94–103
14. Asimov I (1985) Robots and empire. Doubleday, New York

Mario De Caro is a Professor of Moral Philosophy at Roma Tre University and a Visiting Professor at Tufts University, where he has taught since 2000. A former Fulbright Fellow at Harvard University and a Visiting Scholar at MIT, he has also served as president of both the Italian Society of Analytic Philosophy and the Italian Society of Moral Philosophy. He is an associate editor of the *Journal of the American Philosophical Association*, a member of the Steering Committee of the International Federation of Philosophical Associations (FISP), and the literary executor of Hilary Putnam. He has delivered lectures at over 100 universities in 19 countries and published more than 200 scholarly articles in six languages. He is the author of eight books in Italian and the editor or co-editor of over 20 volumes in English and Italian, including *Naturalism in Question* (Harvard University Press), *Naturalism and Normativity* (Columbia University Press), *Practical Wisdom* (Routledge), *The Routledge Handbook of Liberal Naturalism* (Routledge), and of three volumes of papers by Hilary Putnam (Harvard University Press). His forthcoming books include *Liberal Naturalism* (Harvard University Press) and *Hilary Putnam* (with Maria Baghramian, Polity Press). He is a proponent of the widely discussed view of "Liberal Naturalism" and of the "Aretai model" in virtue ethics. His research interests include moral philosophy, the ethics of AI, philosophy of mind, metaphysics, the history of early modern philosophy and science, and metaphilosophy.

Personal Autonomy and Autonomous Systems: Sameness and Difference

6

Fiorella Battaglia

Abstract

This chapter aims to clarify the transformative effects on the concept of personal autonomy resulting from extending the notion to Artificial Intelligence (AI) technologies, particularly in the defense domain. The complexity of this task is increased by the fact that philosophers have offered a wide range of competing accounts of the autonomous agent's special relation to their own desires and values. I will draw on the concept of autonomy implied in John Martin Fisher and Mark Ravizza's control theory, which, while explaining the theory and practice of autonomy, also identifies two conditions that cannot be appropriately satisfied by autonomous systems, but only by persons who interact and communicate with one another. Despite the superficial sameness between personal autonomy and autonomous systems, control theory allows us to pinpoint an important distance in their underlying features. The seeming similarity masks indeed a profound conceptual divergence between autonomy as a condition of moral agency and autonomy understood as the characterization of a system's operational capacities. This supports, at least in principle, the moral argument that entrusting machines with authority over human life and death is ethically wrong, regardless of how advanced the technology might be.

Keywords

Autonomy · Meaningful human control · Kant · Reactive attitudes and reasons · Epistemic injustice

F. Battaglia (✉)
Dipartimento di Studi Umanistici, University of Salento, Lecce, Italy

Ludwig Maximilians University of Munich, Munich, Germany
e-mail: fiorella.battaglia@unisalento.it

M. Anzalone et al. (eds.), *Reconfiguring Human Autonomy*, Integrated Science 40,
https://doi.org/10.1007/978-3-032-14537-6_6

1 Introduction

The concept of autonomy is key to understanding our relationship with Artificial Intelligence (AI). Autonomy is not merely one of many ethical concerns in the field of AI—rather, "it is the central philosophical problem when we confront the human-machine mutual connection". How we understand and apply the concept of autonomy in relation to machines shapes the entire ethical landscape surrounding AI. In this chapter, I argue that a precise conceptual analysis is needed to avoid misunderstanding of the term. By clarifying what autonomy truly entails—and what it does not—I aim to challenge assumptions about so-called "autonomous" systems and to provide a foundation for a more critical and philosophically informed debate on both the autonomy of machines and that of humans. A key requirement of autonomy is the control condition [1–3]. The control condition modifies our understanding of autonomy by emphasizing the normative dimensions of agency and responsibility. Therefore, the idea of control affects how we understand autonomy and—as a consequence—it is crucial for understanding our relationship with AI. Contrary to views that treat control as a secondary or derivative feature of agency, the concept was introduced precisely because autonomy alone may be insufficient for grounding responsibility. Without a coherent account of control, claims about moral or legal responsibility risk collapsing into uncertainty [4]. This notion loses its technicality and takes on a profile that differentiates human and machine control. It helps to disambiguate the concept of autonomy and therefore to make its profiles clear. Therefore, it can be said that control clarifies the morally relevant dimensions of autonomy: accountability, answerability, and responsiveness to reasons. The concept of control has undergone significant refinement and has emerged as the more nuanced concept of *meaningful human control* (MHC), capturing the essential role of human agency in overseeing autonomous systems [5, 6]. When articulated in terms of MHC, it strikes at the core of what renders AI ethically controversial.

2 Meaningful Human Control (MHC)

Importantly, the concept of meaningful human control (MHC) has not only shaped the debate on lethal autonomous weapon systems (LAWS) but has also proven indispensable in critically assessing the ethical challenges posed by other autonomous technologies, including self-driving cars and surgical robots. This expansion highlights MHC's essential role in defining the moral boundaries of delegating control to machines in various sensitive areas of human action. Concerns about control offer a broad framework for understanding the ethical challenges associated with AI, while also highlighting the unique issues that these technologies introduce. An autonomous car crash causing a passenger's death is perceived by the media and public not as a routine incident, but as an ethically significant event. Somehow, since AI is involved, the incident is supposed to have greater moral significance. While the outcomes are the same, people tend to perceive moral significance differently if the controlling agent is a machine rather than a human.

I believe that the beautiful yet unsettling figure of Maria—an image that captures the essence of the uncanny valley phenomenon [7], where a machine appears eerily "almost but not quite human" and evokes profound discomfort—can help explain the unease we experience when confronted with a lethal machine. There is a pre-reflective, intuitive grasp of the distinction between us and other entities that either appear human-like or act in ways that mimic human behavior. This immediate response plays a central role in shaping how we ascribe moral status and interpret the ethical significance of relational practices. In the case of humanoid machines or highly sophisticated AI, this recognition often manifests as a form of discomfort—a reaction aligned with what has been described as the uncanny valley. What unsettles us is not simply that these entities resemble or act as humans, but that they disrupt the structure of mutual recognition on which our ethical practices unfold. As such, our moral unease does not stem merely from novelty or unfamiliarity, but from a more profound disruption of the second-personal structures—of address, response, and accountability—that underlie ethical life [8–10].

This is exactly the reason why the debate on control [2, 11, 12] has shifted from concerns about control in general to a critical focus on MHC. Before the rise of autonomous systems, discussions naturally focused on forms of human control, making it unnecessary to specify the nature of the controlling agent. However, with the advent of a mixed population of agents—both human and artificial—the need has arisen to explicitly distinguish the type of agent exerting control. The concept of MHC emerged as a critical response to the ethical challenges posed by LAWS. The debate about MHC in the context of LAWS has centered on the crucial question of whether such systems should be granted control over life-and-death decisions.

The notion that a machine could autonomously select and lethally harm human targets without human oversight has faced strong objections from a broad range of scholars [13–15]. However, it is important to note that some scholars counter this position by arguing that it would be unethical to forgo the use of such technologies, particularly when they have the potential to reduce human casualties and increase precision in combat [16, 17]. This debate highlights the complex ethical issues surrounding lethal autonomous weapon systems, [18] and underscores the need for a careful consideration of control, responsibility, and humanitarian outcomes. Amoroso and Tamburrini in their 2020 paper argue for a threefold role for human control on weapon systems to be "meaningful" and explain that this would allow humans to have a more active and influential role in the use of weapons [14]. First, humans must occupy the critical role of 'failsafe actor,' tasked with intervening to prevent or mitigate harmful malfunctions in autonomous weapon systems. This responsibility underscores the indispensable need for human oversight to safeguard against unpredictable errors and ethical breaches, reinforcing the argument that MHC is essential for maintaining accountability and moral legitimacy in the deployment of such technologies. Second, human control must be structured in a way that ensures that responsibility can be clearly attributed, thereby preventing accountability gaps—situations in which it becomes unclear who is responsible for wrongful actions [19]. Such gaps arise when autonomous systems, which lack moral agency and legal personhood, make decisions independently, making it difficult or impossible to hold any human or institution accountable [20]. Addressing these gaps is crucial to maintaining the integrity of legal and ethical frameworks and ensuring that wrongful acts do not go unaccountable. Third, human control must guarantee that decisions affecting the life, physical integrity, and property of individuals remain firmly in human hands rather than being delegated to machines. This is seen as ethically crucial as such decisions carry profound moral weight and require judgment, empathy, and accountability, which are not to be found in machines [21]. Moreover, international legal frameworks, including humanitarian and human rights law, emphasize the need for human responsibility to ensure that the use of force complies with principles of proportionality, distinction, and necessity. Delegating these decisions to autonomous systems risks eroding these safeguards, creating ethical blind spots and legal ambiguities that could lead to unjust outcomes and violations of fundamental rights.

3 From a More Technical Standpoint: Navigating Deontological and Consequentialist Ethics

As previously mentioned, Arkin adopts a contrasting stance, arguing that it would be unethical to forgo the use of autonomous systems [16, 17]. His position is supported primarily from a technical standpoint. According to him, the primary reasons for employing autonomous, robotic, or unmanned systems on the battlefield include the following. First, autonomous and robotic systems enable a smaller number of soldiers to accomplish missions that previously required significantly larger forces. By enhancing individual soldier capabilities, these technologies effectively increase military effectiveness and operational efficiency. Second, robotic systems enable combat operations to be conducted over significantly larger areas than was previously feasible. Third, robotics empower individual soldiers to extend their reach deeper into the battlespace, for instance, by enabling them to observe or engage targets at greater distances. Finally, robotic systems allow for the reduction or removal of soldiers from direct combat roles.

After having presented the theoretical reasons, Arkin moves on to offer four compelling justifications for the deployment of autonomous systems—emphasizing the alleged benefits they promise. First, autonomous systems can potentially reduce human casualties by taking soldiers out of direct harm. Second, they may increase the precision of targeting, thereby minimizing collateral damage and civilian deaths. Third, such systems can operate tirelessly and with consistent effectiveness, unlike humans who may suffer from fatigue or emotional stress. Finally, autonomous systems could enhance compliance with the laws of armed conflict by making more calculated and unbiased decisions based on pre-programmed ethical guidelines. Although these points possess a certain degree of argumentative force, LAWS have faced many criticisms and objections in both public and academic discourse. In this regard, there are three primary ethical objections [19]. First and from a more empirical standpoint, based on our current understanding of technological development, robots in the foreseeable future will not possess the capacity to make the complex practical and moral distinctions required by the laws of armed conflict—such as distinguishing between combatants and non-combatants, assessing proportionality in the use of force, and determining the military necessity of violent action. As a result, delegating military tasks to autonomous systems risks increasing the occurrence of wrongful acts and violations of international humanitarian law during military operations. Second, on more principled grounds, it is morally wrong to allow a machine to exercise control over decisions concerning the life and death of human beings, regardless of how advanced the technology may be. According to this position, such applications are considered *mala in se*: intrinsically wrong by their very nature, irrespective of context or outcome. This stance rests on the belief that certain moral decisions, particularly those involving the taking of human life, must remain within the realm of human agency, where accountability, empathy, and moral deliberation are possible. Delegating such decisions to autonomous systems, which lack consciousness, intentionality, and moral responsibility, constitutes a fundamental violation of human dignity. The two positions reflect a deeper ethical divide, with one rooted in deontological principles—emphasizing duties and moral absolutes—and the other grounded in consequentialist reasoning, focusing on outcomes and overall harm reduction. From a deontological perspective, allowing machines to make life-and-death decisions undermines the moral status of persons by treating them as objects of calculation rather than subjects of moral concern. Even if such systems could act more precisely or efficiently, the act itself remains ethically impermissible. This view stands in contrast to consequentialist arguments—such as those advanced by Ronald Arkin—which focus on outcomes, suggesting that if autonomous systems can reduce harm or increase compliance with the laws of war, their use may be not only permissible but also morally required [17, 22].

The debate, then, hinges on whether moral permissibility is determined solely by outcomes or whether certain roles, especially those involving lethal force, are inherently reserved for moral agents. From a more theoretical standpoint, the involvement of autonomous weapon systems in military operations may overcomplicate, or even render impossible, the attribution of moral and legal responsibility in cases of war crimes or fatal accidents [19]. In recent years, the ethical concerns surrounding autonomous weapon systems have been distilled into a guiding principle within the legal-political debate, as reflected in Article 36 [23]. The Principle of MHC asserts that future weapon systems must ensure that humans, rather than computers or algorithms, retain ultimate control over the use of (lethal) force [15]. This means that human agents should remain morally responsible for all significant decisions related to (lethal) military operations. Given the central conceptual outcome of the chapter, the following sections aim to make this insight more intelligible through practical elaboration and contextual application. In particular, to render the principle of MHC operational, it is necessary to articulate its requirements from both technical and legal perspectives. Only by grounding the principle in concrete mechanisms of system design and established legal norms can it effectively guide the development and governance of autonomous technologies.

4 Moral Theory, Control and Autonomy

Santoni de Sio and van den Hoven [5] provide an analysis of the concept of "control" grounded in the philosophical literature on free will and moral responsibility, where control is a central theme. Their perspective is shaped by the longstanding debate in moral theory concerning compatibilism and incompatibilism. Compatibilism and incompatibilism no longer play a central role in moral theory as they once did. Nevertheless, it is important to acknowledge that the paper by Santoni de Sio and van den Hoven contributes to the so-called compatibilist theory of moral responsibility, which holds that individuals can be held accountable even if their actions are causally determined, potentially within a deterministic framework. More importantly, the paper demonstrates how MHC can be systematically designed for, thereby extending the value-sensitive design approach within the ethics of technology to a new and critical domain. There are alternative approaches that, while strengthening certain aspects of MHC, remain insufficient to fully meet the comprehensive set of requirements necessary for MHC to be adequately ensured. Being physically or operationally present in the control loop does not, in itself, ensure that a human is meaningfully guiding or responsible for the system's actions. This mere proximity to the system is insufficient, as a person may be present and able to influence certain aspects of the system through causal intervention, while remaining unable to affect other components—some of which may be more ethically or operationally significant. Even exercising control in the form of making a substantive causal contribution may be insufficient to qualify as MHC. As a matter of fact, the conditions for MHC closely resemble the most stringent standards for legal responsibility. Santoni de Sio and van den Hoven argue that agents must not only possess causal control over events but also satisfy stricter conditions related to knowledge, intention, capacity, and opportunity.

The account of control—and, connected with it, of autonomy—can be situated on a metaethical level [24]. Incompatibilists hold that causal explanations are fundamentally incompatible with genuine autonomy and moral responsibility. This view is reflected both in theories that posit a metaphysical realm of freedom to safeguard the notion of free will and in those that reject free will altogether by extending causal determinism to all events. Contemporary libertarians (e.g., [25–27]), following a philosophical tradition dating back at least to Immanuel Kant, maintain that certain human beings possess a distinctive form of autonomy often referred to as "contra-causal" freedom. This capacity grounds moral responsibility and autonomy in a unique metaphysical status, establishing them in a way that no other kind of being can plausibly claim. From a metaethical perspective, compatibilists believe that humans can be morally responsible for some of their actions, even if they lack any special metaphysical power to escape the influences that cause their behavior. Recent developments in compatibilist thought [1, 2, 7, 28], adopt a more nuanced view of the human mind and agency. They reject the notion that mere mental causation is sufficient to ground autonomy and moral responsibility. Instead, they emphasize the importance of an agent's capacity for rational control over their actions—specifically, the ability to reflect upon, evaluate, and act in accordance with reasons—as the foundation of moral responsibility. There is some kind of sympathetic alignment with current accounts, such as Strawson's [8] and Darwall's [9, 10], highlighting the ways in which our moral practices are deeply rooted in interpersonal relationships and social expectations. People often perform actions that others may not consider morally justified [8, 29]. When challenged, individuals typically respond by offering reasons they believe justify their actions. In doing so, they engage in a process of assuming and negotiating responsibility. Importantly, this process is not isolated or purely internal—it is socially situated. In sum, responsibility is embedded in (1) the roles individuals occupy within social institutions, (2) the continuity of their past actions and character, and (3) the cultural norms and moral expectations that define what counts as acceptable behavior. Similarly, autonomy, intrinsically bound to the notion of responsibility, resists reduction to the detached frameworks typical of purely technical interpretations.

Ultimately, these considerations highlight how Arkin's account, although technically robust, fails to address the more nuanced aspects of the relationship between autonomy and responsibility that make the ethical design of autonomous systems inseparable from the concept of meaningful human control. Consequently, the proposal to treat the deployment of unmanned systems on the battlefield as a kind of categorical imperative is met with an equally substantive and coherent counterargument.

5 The Second-Person Standpoint

This relational view of responsibility aligns with what Darwall [9] has called the second-person standpoint: an ethical framework that sees moral obligations as arising not from detached judgment, but from interpersonal demands made between moral agents. Responsibility, in this view, presupposes the capacity for mutual recognition, communication, and answerability—capacities that machines do not possess. This has direct implications for debates on artificial agency and accountability. Unlike humans, autonomous systems cannot participate in second-person moral practices: they cannot recognize others as moral equals, respond to blame with justification or remorse, or internalize shared norms. As a result, attributing moral responsibility to machines risks bypassing the deeply relational structure of human accountability. If responsibility presupposes this kind of mutual moral engagement, then current AI systems—no matter how advanced—lack the requisite autonomy for bearing moral responsibility in the same sense as human agents. Robert Brandom's inferentialist philosophy, particularly developed in *Making It Explicit* [30], argues that the ability to participate in practices of giving and asking for reasons is foundational to what it means to be an autonomous, responsible agent. For Brandom, meaning is not simply a matter of representing the world, but of being embedded in social practices where claims are made, challenged, justified, and assessed. This discursive interaction defines our status as sapient beings—creatures who can not only follow norms but also justify their actions and beliefs by appealing to shared standards. Brandom directly links this practice to responsibility, arguing that to assert something or to act within a normative space is to undertake a commitment and accept accountability to justify it. The key aspect in his argument is that the capacity to enter justificatory relations is not merely cognitive or computational—it is social and normative. One must be able to recognize others as reason-givers and reason-demanders, and in turn be recognized by them.

This has profound implications when applied to the ethics of artificial intelligence and autonomy. AI systems, no matter how advanced, do not participate in these normative discursive practices. They do not take responsibility for their assertions or actions as Brandom understands that, because they do not recognize themselves—or others—as participants in a community of practice. They process inputs and produce outputs, but they do not enter commitments, respond to challenges, or modify their positions based on mutual understanding or moral reflection. From this standpoint, it becomes clear that autonomy in the human, normative sense cannot be attributed to machines simply because they operate independently. Without the capacity for taking and giving reasons, they lack the core of what Brandom identifies as autonomous agency—and by extension, they cannot bear moral responsibility in any robust sense [31]. Brandom's emphasis on the normative practice of giving and asking for reasons underscores a fundamental limitation of autonomous systems in ethical decision-making. Clearly, machines cannot enter the mutual relationships of recognition and accountability that confer moral significance on decisions. It follows that MHC requires humans to retain ultimate authority and responsibility over decisions involving life, death, and moral judgment precisely because humans are the only agents capable of participating in these norm-governed, discursive practices. MHC thus reflects the necessity of preserving a second-person standpoint—a relational context where agents hold one another accountable through reasons, justifications, and mutual recognition. Machines, lacking this capacity to engage in normative discourse, cannot assume the role of responsible agents. They cannot justify their actions or respond to moral demands; instead, they function as causal devices devoid of moral accountability.

This framework challenges that idea of machine ethics which aims to design AI systems that can make ethical decisions autonomously. Without the capacity for genuine reason-giving and taking, machine ethics remains a form of simulated morality, a set of programmed rules and responses rather than authentic moral agency. Consequently, the ethical control of AI systems must remain in human hands—not only to ensure responsibility but also because moral autonomy is intrinsically tied to the human capacity for normative engagement. In other words, Brandom's framework justifies the insistence on MHC as a conceptual and moral necessity: machines can assist or execute certain functions, but ultimate control and moral responsibility must rest with human beings who can participate in the dialogue of reasons that underpins ethical life.

Building on this, I intend to introduce another account that supports a second-person standpoint. While developed within the basic conceptual structure of the relationship between philosophy and neuroscience, the account by Ravizza and

Fisher is nevertheless capable of effectively engaging in the machine ethics debate. Their framework emphasizes the relational and normative dimensions of autonomy, making it particularly relevant for distinguishing between human autonomy and the autonomy exhibited by artificial systems. According to Fischer and Ravizza [2], in order for a person to be morally responsible for an action X, they must possess what the authors call "guidance control" over that action. This concept implicitly conveys a robust notion of autonomy, as it requires not only the capacity to respond to reasons but also a reflective identification with the mechanism by which decisions are made. The concept of guidance control, as developed by Fischer and Ravizza, provides a substantive account of what it means to act autonomously in a morally responsible way. Autonomy, in this context, is not simply a matter of acting independently or without external coercion; rather, it involves acting through a decision-making mechanism that is both moderately reason-responsive and owned by the agent. The first condition—moderate reason-responsiveness—links autonomy to the agent's capacity to recognize, evaluate, and act upon moral and practical reasons. This ensures that the agent's behavior is not merely the product of blind causation or internal compulsion but reflects responsiveness to normative considerations and relational practice. The second condition—that the mechanism be "the agent's own"—introduces a deeper, relational dimension to autonomy. For Fischer and Ravizza, an agent must not only possess a functional capacity for responsiveness but must also have taken ownership of the process through which decisions are made. This involves understanding how one's actions are generated and accepting responsibility for them, which in turn requires a degree of self-reflection and integration into a broader moral community.

Taken together, these conditions suggest that autonomy, properly understood, is not just a matter of control or choice, but of normative engagement and identification with one's reasons for acting. This conception contrasts sharply with simplified notions of autonomy often attributed to machines, where autonomy is equated with independent functioning rather than with moral responsibility or reflective ownership. Darwall [9] argues that moral obligation, accountability, respect for persons, and the distinctive freedom of moral agents are all irreducibly second-personal or relational. This means that these concepts cannot be fully understood from a detached, third-person perspective [32–34] or even a purely first-person point of view [35]. Instead, they arise within a space of reciprocal moral address, where individuals make claims, demands, and justifications to one another as equals in the moral community. From this standpoint, to say someone is morally obligated is not merely to describe their behavior or attitudes, but also to state that others have standing to hold them accountable, to demand reasons for their actions, and to expect certain forms of acknowledgment or redress. Responsibility, then, is not a solitary function of inner conscience or rational deliberation alone—it depends on the interpersonal structure of recognition and answerability.

6 Relational Autonomy

Some significant implications for the concept of autonomy can be pinpointed. In Darwall's framework, autonomy is not limited to self-governance or internal coherence; it also involves the capacity to engage in mutual moral relations, respond to others' demands, and acknowledge the authority of shared moral norms. Autonomy, in this sense, is inherently relational and dialogical. Applied to contemporary debates in AI ethics, this perspective underscores that autonomous systems, no matter how sophisticated, are fundamentally incapable of genuine second-personal engagement. They do not recognize others as moral equals, nor can they issue or respond to moral demands in a way that reflects genuine accountability or respect. As such, they cannot be full participants in moral practices grounded in Darwall's second-person standpoint—reinforcing the argument that moral responsibility and autonomy remain uniquely human capacities. Brandom, Strawson, and Darwall collectively support a conception of autonomy that is not simply about individual self-governance or independence, but instead it emphasizes the importance of our involvement in a normative community where agents hold each other accountable. According to this perspective, autonomy is exercised through participation in practices of giving and asking for reasons [30], being subject to and issuing reactive attitudes like blame or resentment [8] and recognizing oneself and others as bearers of moral authority and obligations [9]. Relational theorists of autonomy argue that appropriate social relationships are something more than just supportive of autonomy: they are fundamental to its very structure.

Following this line of thought, autonomy should not be conceived merely as internal self-governance, but rather as a capacity that is cultivated, acknowledged, and sustained within a network of social and interpersonal relationships. Scholars such as Oshana [36], Mackenzie [37], and Nedelsky [38] emphasize that autonomy requires conditions of mutual recognition, respect, and access to genuinely meaningful choices. Without these, even agents who meet internal criteria may lack the standing or opportunity to act autonomously in any robust sense [39]. In this light, being appropriately related to others—being heard, taken seriously, and afforded fair options—is not a causal condition that helps us to interact with others, but rather a pivotal element for autonomous agency. Such a relational perspective challenges the idea of the autonomous individual as isolated and self-contained, and instead places autonomy within the very fabric of social life. This insight into the relational foundations of autonomy resonates with accounts of language and normativity that similarly recognize the social character of agency. In this regard, Brandom defends a conception of discursive practice in which language is structured normatively through what he calls its "downtown": the game of giving and asking for reasons. At the heart of this practice is the idea that to speak is to engage in a normative activity—advancing claims, drawing inferences, and navigating a network of commitments and entitlements. When speakers make assertions, they undertake commitments that can be challenged or defended through reasons. If successfully justified, these commitments earn entitlement. Importantly, this is not just one language game among many; rather, it is constitutive of discursive practice itself.

Without this normative structure, there would be no meaningful linguistic communication at all. Rather than being a solitary or inward capacity, autonomy emerges through second-personal engagement which unfolds as recognizing others as sources of legitimate moral claims and being recognized in turn.

The idea of relational autonomy, articulated in this way, effectively demonstrates the radical difference between human autonomy and machine autonomy—a difference that this chapter has sought to highlight. It also shows that the moral significance of our actions is inseparable from the shared practices, expectations, and mutual recognition that constitute our social world. Autonomy gains its depth and normative force precisely because it is embedded in the meaningful webs of interpersonal connection—and machines cannot impose themselves as leading agents within this uniquely human sphere. Recognizing this fundamental distinction has important implications not only for the design of autonomous systems but also for how we conceive of the human-machine relationship itself. As technological capabilities continue to advance, it remains essential to remember that authentic moral agency depends on forms of engagement and accountability no machine can replicate. Although artificial systems may appear increasingly like us in their operations, there are radical differences that must be clearly acknowledged, to avoid human autonomy being conflated with artificial autonomy—and ultimately risk being overshadowed by it.

7 Conclusion

The conception I have been arguing for situates autonomy within a relational and dialogical space, where moral agency depends on responsiveness to norms that are shared, upheld, and negotiated between persons. It does not reject Kant's emphasis on autonomy as a necessary presupposition of moral obligation and the dignity of persons. However, it offers a distinct interpretation of this idea by emphasizing that these moral concepts contain an irreducibly second-personal dimension. In this view, autonomy is not simply the capacity for self-legislation of a rational will but is grounded in our ability to relate to others as mutually accountable moral agents—agents who have reactive attitudes and can issue, recognize, and respond to moral claims within a shared normative space. This sharply contrasts with the algorithmic or operational autonomy of artificial systems, which lacks the relational dimension of human autonomy. Machines may act independently, but they do not possess the capacity to recognize or respond to the moral authority of others. As such, they cannot be autonomous in the second-personal sense, nor can they bear the moral responsibilities that follow from such autonomy.

Competing Interests This work was partially supported by the project "Ethical Design for AI", part of the Spoke 6 of the project FAIR–Future AI Research (PE00000013), under the NRRP MUR program funded by the NextGenerationEU and by the European Union–Next Generation EU, Missione 4 Componente 1 CUP F53D23010740001 DigitHuman.

References

1. Dennett DC (1984) Elbow room: the varieties of free will worth wanting. MIT Press, London
2. Fischer JM, Ravizza M (1998) Responsibility and control: a theory of moral responsibility. Cambridge University Press, New York
3. Shepherd J (2015) Deciding as intentional action: control over decisions. Australas J Philos 93(2):335–351
4. Di Nucci E, Santoni de Sio F (2014) Who's afraid of robots? Fear of automation and the ideal of direct control. In: Battaglia F, Weidenfeld N (eds) Roboethics in film. RoboLaw series. Pisa, Pisa University Press, pp 127–144
5. Santoni de Sio F, van den Hoven J (2018) Meaningful human control over autonomous systems: a philosophical account. Front Rob AI 5:1–14
6. Robbins S (2024) The many meanings of meaningful human control. AI Ethics 4:1377–1388. https://doi.org/10.1007/s43681-023-00320-6
7. Dennett DC (2014) When AI kills who is to blame? In: Battaglia F, Mukerji N, Nida-Rümelin J (eds) Rethinking responsibility in science and technology. Pisa University Press, Pisa, pp 203–214
8. Strawson P (1963) Freedom and resentment. Proc Br Acad 48:187–211
9. Darwall SL (2006) The second-person standpoint. Harvard University Press, Cambridge
10. Darwall SL, Dill B (2014) Moral psychology as accountability. In: D'Arms J, Jacobson D (eds) Moral psychology and human agency: philosophical essays on the science of ethics. Oxford University Press, New York, pp 40–83
11. Mele AR (1995) Autonomous agents: from self-control to autonomy. Oxford University Press, New York
12. Greenspan P (1978) Behavior control and freedom of action. Philos Rev 87:225–240
13. Asaro P (2012) On banning autonomous weapon systems: human rights, automation, and the dehumanization of lethal decision-making. Int Rev Red Cross 94(886):687–709. https://doi.org/10.1017/S1816383112000768
14. Amoroso D, Tamburrini G (2020) Autonomous weapons systems and meaningful human control: ethical and legal issues. Curr Robot Rep 1:187–194
15. Véliz C (2021) Moral zombies: why algorithms are not moral agents. AI & Soc. https://doi.org/10.1007/s00146-021-01189-x
16. Arkin R (2008) Governing lethal behavior: embedding ethics in a hybrid deliberative/reactive robot architecture. In: Proceedings of the 3rd ACM/IEEE international conference on human–robot interaction. Association for Computing Machinery, New York, pp 121–128. https://doi.org/10.1145/1349822.1349839
17. Arkin R, Ulam P, Wagner AR (2012) Moral decision making in autonomous systems: enforcement, moral emotions, dignity, trust, and deception. Proc IEEE 100:571–589
18. Blanchard A, Chris T, Taddeo M (2023) Ethical governance of artificial intelligence for defence: normative tradeoffs for principle to practice guidance. Available at SSRN: https://ssrn.com/abstract=4517701 orhttps://doi.org/10.2139/ssrn.4517701
19. Santoni de Sio F, Mecacci G (2021) Four responsibility gaps with artificial intelligence: why they matter and how to address them. Philos Technol 34:1057–1084. https://doi.org/10.1007/s13347-021-00450-x
20. List C, Pettit P (2011) Group agency: the possibility, design, and status of corporate agents. Oxford University Press, Oxford/New York
21. Yıldız T (2025) The minds we make: a philosophical inquiry into theory of mind and artificial intelligence. Integr Psychol Behav Sci 59:10. https://doi.org/10.1007/s12124-024-09876-2
22. Wagner A, Borenstein J, Arkin R (2019) Kantian one day, consequentialist the next: moral emotions as mediators between ethical frameworks for robots. In: Proceedings of the International Association for Computing and Philosophy (IACAP), June 2019

23. Art 36 (2015) Killing by machine: key issues for understanding meaningful human control. http://www.article36.org/autonomous-weapons/killing-by-machine-key-issues-for-understanding-meaningful-human-control/
24. Abbink D, Amoroso D, Cavalcante Siebert L, van den Hoven J, Mecacci G, Santoni de Sio F (2024) Introduction to meaningful human control of artificially intelligent systems. In: Research handbook on meaningful human control of artificial intelligence systems. Edward Elgar Publishing, Cheltenham, pp 1–11
25. van Inwagen P (1983) An essay on free will. Oxford University Press, New York
26. Kane R (1996) The significance of free will. Oxford University Press, New York
27. Hodgson D (2012) Rationality + consciousness = free will. Oxford University Press, New York
28. Frankfurt HG (1971) Freedom of the will and the concept of a person. J Philos 68(1):5–20. https://doi.org/10.2307/2024717
29. Korsgaard CM (1992) Creating the kingdom of ends: reciprocity and responsibility in personal relations. Philos Perspect 6:305–332. https://doi.org/10.2307/2214250
30. Brandom R (1994) Making it explicit: reasoning, representing, and discursive commitment. Harvard University Press, Cambridge, MA
31. Baum K, Mantel S, Schmidt E et al (2022) From responsibility to reason-giving explainable artificial intelligence. Philos Technol 35:12. https://doi.org/10.1007/s13347-022-00510-w
32. Snow CP (1959) The two cultures and the scientific revolution. Rede lecture. Cambridge University Press, Cambridge
33. Pereboom D (2001) Living without free will. Cambridge University Press, New York
34. Greene J, Cohen JD (2004) For the law, neuroscience changes nothing and everything. Philos Trans R Soc B 359(1451):1775–1785
35. Nagel T (1986) The view from nowhere. Oxford University Press, New York
36. Oshana M (2006) Personal autonomy in society. Ashgate, Aldershot
37. Mackenzie C (2000) Imagining oneself otherwise!: toward a feminist theory of relational autonomy. In: Mackenzie C, Stoljar N (eds) Relational autonomy: feminist perspectives on autonomy, agency, and the social self. Oxford University Press, New York. https://doi.org/10.1093/oso/9780195123333.003.0007
38. Nedelsky J (2011) Law's relations: a relational theory of self, autonomy, and law. Oxford University Press, Oxford
39. Fricker M (2007) Epistemic injustice: power and the ethics of knowing, Online edn. Oxford Academic. https://doi.org/10.1093/acprof:oso/9780198237907.001.0001. Accessed 6 July 2025

Fiorella Battaglia is the Founding Director of the Laboratory for Ethics in the Wild at the Digital Humanities Centre, University of Salento, where she is also associate professor of moral philosophy in the Department of Humanities. Her research focuses on challenging ethical questions resulting from emerging technologies and climate change, which shape both our social and epistemic practices and our moral experiences. After obtaining her MA degree in Philosophy from the University of Pisa, she earned her PhD in Philosophy and Politics from the University of Naples "L'Orientale" (2004). In 2016, she completed her habilitation in Practical Philosophy and received her venia legendi from the Ludwig-Maximilians-Universität in Munich (Germany). She has also held an assistant professorship of Social Philosophy at the Berlin-Brandenburg Academy of Sciences and Humanities, at the Humboldt University in Berlin, an adjunct professorship of Epistemology at the Faculty of Medicine of the University of Pisa, and a visiting professorship at the Dirpolis and Biorobotics Institutes of the Sant'Anna School of Advanced Studies in Pisa.

Autonomy-Enhancing Technologies? Automated Vehicles and Autonomous Mobility

7

Fabio Fossa

Abstract

Automated Vehicles (AVs) are often showcased as technologies that will considerably enhance the autonomy of their users. Indeed, mobility enables people to better satisfy crucial needs and pursue self-determined goals, thus representing a practical condition for the exercise of human autonomy. While manual driving excludes many (often vulnerable) social groups such as the disabled, the elderly, and the young, automated driving would include them among its users. Moreover, the delegation of driving tasks to AI systems would liberate travellers from the burden of driving, thus freeing time and energy for the pursuit of individual goals. However, the claim that autonomous mobility would be fostered by driving automation *per se* is a simplification that rests on a misconception of the "autonomy" of AVs. By criticizing techno-solutionist arguments presenting AVs as autonomy-enhancing technologies, this chapter aims at stressing the irreducible socio-political nature of the challenge of supporting human autonomy through transport innovation.

Keywords

Ethics of technology · Autonomy · Automated vehicles · Mediation · Transportation ethics

F. Fossa (✉)
Department of Mechanical Engineering, Politecnico di Milano, Milan, Italy
e-mail: fabio.fossa@polimi.it

M. Anzalone et al. (eds.), *Reconfiguring Human Autonomy*, Integrated Science 40,
https://doi.org/10.1007/978-3-032-14537-6_7

1 Introduction

Several Artificial Intelligence (AI) technologies appear to generate high expectations, masterfully fuelled by the companies that develop them, regarding the positive impacts they might turn out to have on human lives and the conditions of the planet. In addition to practical and economic values, the hype surrounding them often emphasizes distinctively moral values as well. In this way, the worth of AI technologies is also justified by appealing to their supposed *goodness*—i.e., their potential to improve morally unsatisfactory conditions and bring about ethically relevant benefits.

In the transportation field, such a blend of commercial and ethical motives visibly stands out when driving automation is discussed. According to a narrative that is now widespread in both business and academic circles [1], Automated Vehicles (AVs) are expected to yield nothing less than a moral revolution in transport. Thanks to the delegation of driving to AI systems—so it is argued—mobility needs driven by values like safety, inclusivity, sustainability, and autonomy are bound to be met as never before.

A closer look at the issue soon reveals that the desired goal is much further off than assumed. A revolutionary moral transformation of transportation systems would require much more—and perhaps, especially in the short term, something quite different [2–4]—than a rapid spread of AVs on our road networks. Thus, carrying out a deeper examination of the relationship between driving automation and the moral values it is expected to support can be valuable and instructive not just for debates on the future of transportation, but for AI and technology ethics as well. To this aim, in the following pages I will attempt to clarify how AVs *mediate* ethical values that are relevant in the practical context of transportation.

The philosophy of technological mediation as presented in the post-phenomenological tradition (e.g., [5–10]) is therefore assumed here as the main methodological perspective from which to call into question the claim underlying the hype that technologies such as AVs can be framed as *unambiguously value-enhancing artefacts*. Very briefly, post-phenomenological thinking argues that technological mediation contributes to shaping—i.e., co-shapes—human experience by making some of its aspects easier to perceive or act upon (*amplification*) while making others more difficult to perceive or act upon (*reduction*). This applies to moral experiences as well. In this sense, post-phenomenology suggests addressing the entanglements of ethical values and technologies as instances of a process of mediation where human practices are transformed and their moral profile considerably remodulated. Indeed, technologies co-shape human practices also by amplifying and reducing, or supporting and hindering, the expression of ethical values that characterize them. Therefore, the philosophy of technological mediation invites to look at the multiple, ambiguous ways in which ethical values are conveyed in relations with technologies and stresses the need not to lose awareness of the moral nuances that this process hides, reduces, or hinders. As such, not only it offers an insightful point of view on the intricate issue of the entanglement between ethical values and artefacts, but also provides a fruitful perspective from which to problematize the one-sided view about the future impacts of AVs that is typical of hyped narratives.

The reflections that follow are motivated by three intentions. The first is to stress that possible benefits of AI systems (such as AVs) are rarely straightforward by-products of their adoption. Rather, these benefits can only be reaped if set as explicit goals from the early stages of technology design and societal deployment—which often conflicts with other design and business objectives. Presenting a narrative where moral goals seamlessly align with design and business aims is misleading and warrants critique. The second intention is to emphasize that, even when efforts are made to align AI systems with relevant moral values, challenges remain due to the ambiguity of their technological mediation. The third intention is to suggest that the responsible management of such ambiguity should be addressed by considering the broader practical and normative context to which it belongs, rather than any specific technology that might raise it. In our case, an overly narrow focus on driving automation risks losing sight of the broader mobility context in which the moral challenges that AVs seek to address are rooted. The ethical questions raised by driving automation assume their proper significance only in the context of a wider transportation ethics.

Driven by these intentions, the chapter focuses on the ambiguous ways in which the ethical value of *human autonomy* is mediated through driving automation. The complex relationship between human autonomy and technological automation is central to the debate not only in the case of driving automation, but also of other AI technologies (e.g., [11–13]) such as autonomous weapons systems (e.g., [14, 15]) and recommendation systems (e.g., [16–20]). It could hardly be otherwise. In Western culture, autonomy is widely recognized on a sociopolitical level as a fundamental ethical value (e.g., [21]). The notion of human autonomy can hardly be defined uncontroversially, as it has been characterized in multifarious ways throughout the vast philosophical literature exploring it. Since the present work requires some clarity on what "human autonomy" is taken to mean, but does not allow for lengthy conceptual discussions, the following has to suffice for the moment. In brief and not without simplification, autonomy is here taken to characterize human beings as entities capable of self-determination even though they are subject to various influences and situated within multiple relational contexts. On this basis, humans are believed to deserve respect and protection. Autonomy underpins crucial aspects of the moral life, such as the exercise of responsibility and the full enjoyment of human dignity. Hence, it qualifies as a key value to respect and pursue, also through technological innovation—driving automation included.

2 Hype Alert! Human Autonomy and Autonomous Vehicles

For these same reasons, a technology that can potentially amplify or facilitate the exercise of human autonomy is well-suited to being promoted as *good*. References to autonomy are common in the hype buzzing around AVs, which rarely fails to highlight how driving automation is bound to revolutionize its enjoyment. According to the narrative mentioned above, autonomous driving will allow individuals to reclaim time and energy previously spent on driving, which could then be devoted to freely chosen activities. This will also lead to an increase in decision-making autonomy on issues that significantly affect individual well-being—such as where to live, study, or work—but are currently constrained by the ability to drive and the burden of daily commutes. Moreover, autonomous vehicles will offer a new private transportation option to individuals who currently cannot benefit from it due to cognitive or physical disabilities, age, or personal preferences, significantly improving their ability to control their lives and make independent decisions.

Just as an example, consider the case of Waymo—one of the most advanced companies in the driving automation sector. Waymo's narrative significantly characterises driving automation as freeing individuals from the limitations and burdens that come attached to manual driving. In terms of limitations, driving automation is envisioned as bringing mobility to social categories who cannot access it directly—the young, the elderly, the disabled, those who do not feel comfortable in driving themselves. As co-CEO Tekedra Mawakana explains in an interview at Code 2021 [22], Waymo aims at "really giving people independence", so that "they don't need mom or dad or girlfriend or boyfriend to carry them around" [15:32]. At the same time, it aims at minimizing burdens, especially in the form of avoiding wasting time in driving—time that can be then put to better use according to one's own personal preferences. "We're human", Mawakana states, "and when we get in the car we have other stuff to do" [11:05]. Waymo's vision of driving automation embraces the liberating opportunity of giving "people the chance to just do what they want to do—which is not focus on driving" [11:15]. Everything in Waymo's communicative strategy supports the individual search for freedom through mobility—a core value traditionally integral to private transportation, but somewhat threatened when driving is taken out of human hands and transferred to AI systems.

If one was to take this narrative for granted, it would seem that there is no ambiguity in how driving automation mediates the value of autonomy. The mediation appears to be entirely positive: as claimed, autonomous vehicles inherently amplify the exercise of human autonomy. However, a critical examination of technical and regulatory aspects reveals a more ambiguous situation. Supports and obstacles to the exercise of autonomy vary, are interwoven, and most importantly result from explicit design choices. To the extent that responsible design decisions and effective policies depend on a rigorous understanding of opportunities and risks, a philosophical investigation into the mediation of human autonomy in driving automation is necessary.

To uncover the ambiguity hidden by the hype, a good starting point is the European approach to the ethics of automated and connected driving—an approach that substantially mirrors the central position of the value of autonomy in Western culture. In 2020, an interdisciplinary group of 14 experts appointed by the European Commission drafted the report *Ethics of Connected and Automated Vehicles. Recommendations on Road Safety, Privacy, Fairness, Explainability, and Responsibility* [23]. The document presents moral principles and recommendations aimed at supporting stakeholders in their efforts to align driving automation with the most relevant ethical values. Closely related to the European approach to trustworthy AI [24], the report opens by identifying and defining the moral foundations that inspire the entire proposal. Recognizing its importance, the authors include autonomy among the eight fundamental ethical principles for driving automation, alongside non-maleficence, beneficence, dignity, responsibility, justice, solidarity, and inclusive deliberation.

According to the report, the principle of autonomy asserts that human beings should be conceived as "free moral agents," whose right to self-determination deserves respect. In relation to autonomous driving, the principle of autonomy requires that driving systems be designed to "protect and promote human beings' capacity to decide about their movements and, more generally, to set their own standards and ends for accommodating a variety of conceptions of a 'good life'" [23, p. 22]. Consequently, autonomy plays a crucial role in various recommendations, ranging from privacy protection to promoting users' decision-making capabilities and reducing algorithmic opacity.

From an ethical point of view, the emphasis on protecting and promoting human autonomy in the context of driving automation seems appropriate. Hindering autonomy in decision-making and action through automated driving systems risks engendering situations where crucial aspects of individuals' lives are determined by other agents (e.g., computer scientists, engineers, manufacturers). This state of affairs is clearly incompatible with the individual right to self-determination in personal matters and must be avoided in the design and introduction of AVs into transportation systems.

In addition to requiring intrinsic respect, protecting human autonomy in the context of driving automation is also important in an instrumental sense, i.e., to uphold other relevant moral values. For instance, considering human beings as free moral agents also enables holding them responsible to the extent that they can exercise such freedom. This is a necessary premise for fairly distributing responsibility when harmful consequences arise from the use of AVs (e.g. [25, 26]).

3 Driving Autonomies

If the normative importance of human autonomy for driving automation is unambiguous, the same cannot be said for its application. Here, on the contrary, significant challenges arise. After all, automating the human act of driving involves partially or fully delegating it to AI systems that mediate its execution. As the driver turns into a supervisor or mere passenger, some limitation of individual autonomy is inevitable. And just like in any other case of technological mediation, when driving is mediated through AI systems, human experience is ambiguously transformed. Some of its aspects get amplified or facilitated, while others get reduced or hindered. And while, from an ethical viewpoint, what gets amplified does not necessarily deserve amplification, what is reduced does not necessarily deserves reduction. To the extent that human autonomy is a part of this experience, its exercise will also be transformed accordingly. It is therefore essential to evaluate which aspects of human autonomy are facilitated or hindered—and to what extent—when driving is mediated through AI systems. Thanks to this understanding, design and policy choices might be able to explicitly address amplifications and reductions that pose threats to autonomy as an ethically valuable component of human existence.

The European report's definition of autonomy helps clarify what is at stake in this context. It identifies two relevant aspects of the exercise of autonomy behind the wheel. The first aspect concerns the exercise of individual control over decisions related to one's movements. In the experience of driving, this refers to the ways in which the vehicle moves. According to the well-known model developed by cognitive psychologist Michon [27], this first aspect includes *operational* and *tactical* driving decisions—i.e., those decisions that are related to vehicle control and navigation on the road network. Examples might include when and how to press pedals, turn the steering wheel, use the turn signal, and so on; and when and how to speed up, slow down, or swerve, which speed to maintain, how much space to leave between the vehicle and vulnerable road users like pedestrians, cyclists, or motorcyclists, whether or not to yield or stop at a crosswalk, and whether to make an exception to traffic laws for a higher purpose, among others. In this sense, respecting user autonomy would imply allowing them to exercise some degree of control over operations affecting how the AV moves on the road.

The second aspect, instead, presents a broader scope. It pertains to transportation systems as essential means for pursuing the good life, regardless of the plurality of morally permissible conceptions of it. Indeed, the accessibility to crucial places, services, and institutions (hospitals, schools, shops, social and political centres, recreational venues, parks, and so on) that transportation systems provide to people is widely recognized as a fundamental enabler of well-being and the autonomous pursuit of a self-determined conception of the good life [28]. As Martens [29] thoroughly explains, different levels of accessibility caused by deficient or poorly balanced transport systems strongly reverberate on people's quality of life, motivating questions about fairness and justice that are more and more perceived as integral to the domains of transport and urban planning. Moreover, being profoundly embedded in extant relations of power, wealth, and influence, transportation risks perpetuating historical injustices and deepening gaps between better-off and worse-off social groups instead of contributing to bridging them. In this sense, new transportation technologies make their entrance in systems that are already characterized by a given level of accessibility and issues of exclusion, segregation, and transport poverty—the effects of which on people's quality of life should not be underestimated [3]. This shows the practical importance that transport innovations have as enablers of that aspect of human autonomy that pertains to the individual pursuit of a self-determined conception of the good life.

Returning to Michon's model, the definition now seems to refer to strategic driving decisions, which concern not so much vehicle control in traffic, but rather the purpose for which a vehicle is used in the first place—i.e., decisions belonging to the general endeavour of living well. Decisions of this kind involve, for example, the destinations to reach, the reasons for wanting to reach them, the preferred routes to take, whether to go alone or with others, and so on. Taken together, these mobility decisions both express and contribute to forming different conceptions of what it is necessary for, and what it means, to live a life worth living. In this sense, aligning driving automation with the principle of autonomy would mean designing it as an instrument to support the independent search for a good life.

When we drive, then, we exercise our autonomy on multiple levels: in how we operate the vehicle; in the driving decisions we make at the wheel; in choosing where, when, how, why, and with whom to travel. Driving involves a multifaceted exercise of autonomy, and driving automation cannot but mediate these various exercises in different ways. This complexity is hardly acknowledged in the hype, but it is critical to capture it.

And this is not even the only source of complexity. A second layer of plurality pertains to the multiple types of driving automation that can be designed. Indeed, there isn't just one type of driving automation, but rather multiple configurations. As detailed in the widely adopted document SAEJ3016 (Fig. 7.1), at least six levels of driving automation can be distinguished based on the kind of human-machine cooperation they exhibit. While L0 stands for *no automation*, L1 and L2 describe increasing forms of (*partial*) *driver assistance* in which automated features such as automatic braking or adaptive cruise control are intended to support users, who however remain drivers in the full sense of the word. The domain of proper driving automation, where driving is delegated to the AI system as a whole, begins at L3.

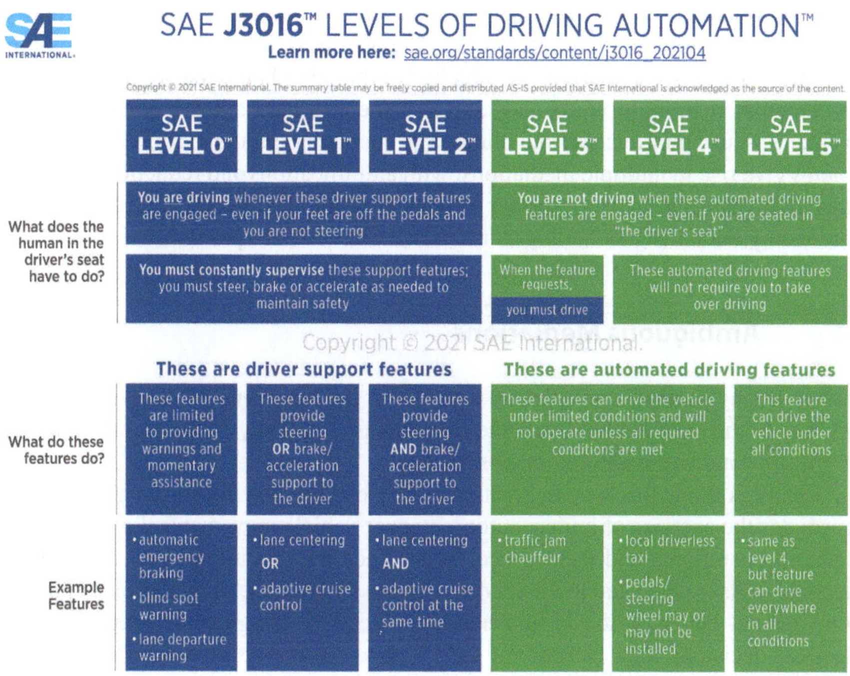

Fig. 7.1 SAE J3016 levels of driving automation [30]https://brxcontent.fullsight.org/site/binaries/content/assets/sae-org/content/news/blog/sae-j3016-visual-chart_5.3.21.pdf

Also known as *conditional automation*, L3 describes features that automate driving only within specified Operational Design Domains (ODDs)—i.e., situations determined by a set of prerequisites including weather, temperature, road, and traffic conditions that specify the circumstances in which L3 systems can be safely activated. L3 requires humans to take over driving both when transitioning between different ODDs and when facing unexpected situations inside a given ODD. Handover and takeover manoeuvres, switching control between humans and systems, are therefore an integral part of conditional automation. Users must always remain vigilant and attentive, occupying an arguably ill-defined role as passengers and drivers at the same time.

L4 (*High Automation*) is envisioned as a form of driving automation that would render takeover and handover manoeuvres mostly unnecessary. It describes systems that are able to operate in given ODDs without requiring users to supervise or intervene. If safety critical situations arise, L4 systems must be able to either handle them satisfactorily or lead the vehicle to a minimal risk condition entirely by themselves, without involving users at all. When similar automated driving features are activated, users can be fully conceptualised as passengers. Finally, Level 5 (*Full Driving Automation*) covers the management of all dynamic driving subtasks on all possible ODDs—i.e., everywhere and under every condition. Only at this level users complete the transformation from drivers to passengers. As it will soon become apparent, different levels of automation facilitate and hinder the different levels of human autonomy exercised in driving in various ways.

To summarize, in the case of both humans and driving systems the notion of autonomy is plural and multifaceted. On the one hand, humans exercise autonomy at different levels when driving a vehicle. On the other hand, the act of driving can be automated in multiple manners, so that the so-called autonomy of automated vehicles is also multifaceted and plural. The encounter of human and vehicle *autonomies* give rise to multiple mediations, with amplifications and reductions, supports and obstacles that call for further ethical scrutiny. The hyped image of AVs as autonomy-enhancing technologies needs to be called into question.

4 Ambiguous Mediations

When the hype curtain is lifted, an ambiguous landscape meets the eye. The complexity of the duty to implement the principle of autonomy in AVs, so unequivocal from a normative standpoint, begins to show.

Let us examine the issue in greater detail. Consider first how driving automation mediates the exercise of human autonomy over vehicle movements, i.e., with reference to operational and tactical decisions. Since the execution of operational and tactical tasks is delegated to the driving system, the exercise of these kinds of autonomy cannot but be variously hindered or reduced. Although some operational and tactical decisions may lack moral significance, others do involve ethical values. Obstacles and reductions of this type arise at every level of driving automation. Even a simple speed control system that does not allow driving beyond the speed limit and cannot be turned off in any situation could be considered an obstacle to the rightful exercise of autonomy [31, 32]. At the opposite extreme, imagine fully autonomous driving systems that in cases of unavoidable collisions could distribute harms based on predetermined ethical values thanks to advanced sensors and control logics. If such values were set not by users but by other parties, their moral autonomy would be hindered or reduced [33, 34]. To consider less futuristic cases, think of the choice to show (or not show) respect and care by driving with extreme caution around vulnerable road users such as cyclists or pedestrians, or the choice to adopt (or not adopt) "eco-friendly" driving styles aimed at reducing noise and emissions. Automating these aspects of driving implies a restriction on the human exercise of autonomy. Its problematic nature becomes evident when vehicle control logics compute driving styles that are different from those that users, if they were in control, would feel obliged to adopt.

In light of the above, it seems reasonable to conclude that the rush toward automation should not hinder forms of human control over driving tasks that allow for the legitimate exercise of autonomy. As suggested by the theory of Meaningful Human Control [35], if certain driving decisions rightfully fall within the purview of user self-determination, AI systems must allow users to make these decisions autonomously. At the level of full automation, where the driving system is explicitly designed not to demand any type of supervision or control by passengers, these requirements would be quite difficult to meet. A minimal form of control could be achieved only indirectly—for example, through the configuration of system settings. However, this solution would arguably fall short of satisfying the demands of the principle of autonomy. The automated application of system settings to unpredictable, complex, and fluid traffic conditions could hardly prevent misalignments between computed driving behaviours and users' contextually self-determined values. It seems more convincing to assert that respect for autonomy understood as control over operational and tactical driving decisions is hindered and reduced by full automation.

In contrast, partial or conditional levels of driving automation, where human drivers and AI systems share control over operational and tactical decisions, seem to leave sufficient room for users to exercise their autonomy. The mediation of their own driving experience, one could argue, does not necessarily reduce or hinder control over operational and tactical decisions. On the contrary, in cases where it offers targeted assistance against distraction, fatigue, stress, or perceptual limitations, partial or conditional automation may even amplify or support the exercise of autonomy behind the wheel by countering constraints to which it is exposed. Consider, for example, a driving system that helps drivers not to exceed a self-determined speed limit even if they become distracted, or a system that signals the presence of cyclists or pedestrians in the vehicle's blind spots and, if necessary, prompts drivers to leave more room or follow safer trajectories displayed on digital car screens. In these instances, technological supports might help human drivers exercise their autonomy at the fullest by mitigating cognitive limitations that constrain its expression while driving.[1]

And yet, the ambiguity remains. The conclusion that human autonomy would be better mediated through partial or conditional levels of automation, where humans and AI systems share driving control, becomes controversial if this value is understood in reference to the second aspect previously identified. Full automation can support the autonomous pursuit of various conceptions of a good life better than partial or conditional automation. At least two opportunities stand out: inclusivity and the reclaiming of travel time. Both significantly and broadly enable the pursuit of different conceptions of the good life, and neither can be realized by relying on partial or conditional automation models.

On the one hand, full automation could significantly enhance the autonomy of social groups currently excluded from manual driving due to physical and cognitive disabilities—individuals for whom independent and private access to transportation has a crucial impact on personal well-being and quality of life. Since driving tasks would be automated, physical and cognitive disabilities would no longer pose an insurmountable barrier to the independent use of private vehicles. This point, which does not escape the hype, is not as straightforward as it seems [38–40]. Individuals with physical or cognitive disabilities will inevitably face other challenges in independently using even the most advanced driving systems unless these systems are fully designed with their peculiar needs in mind. Automating driving is not enough to make it inclusive: specific design choices are needed to address the specific needs of these user groups. An AV designed for the average customer is as exclusive to those who do not fit that category as a manual vehicle is. And as is obvious, designing an AV that is truly inclusive requires commitment and resources far greater than simply distributing the same model to everyone and praising its inclusivity as a by-product. Hype aside, it remains true that full driving automation combined with a design process attentive to the specific needs of given user groups could, at least in principle, offer them a highly significant amplification of autonomy.

On the other hand, full driving automation could facilitate the autonomous pursuit of a good life by allowing (every category of) users not to waste travel time focused on the task of driving. Freed from this burden, AV users could employ travel time as they see fit. Such freedom could expand into further liberating effects on other practical aspects impacting individuals' ability to pursue their own conceptions of the good life, facilitating autonomous decisions on important matters like where to live, work, receive healthcare, and study—all decisions that would be less constrained by the time wasted traveling from one place to another.

[1] Even in this case, however, the mediation is ambiguous. Since these systems highlight risks or suggested trajectories, they are prone to causing attention tunnelling—a psychological effect according to which attention is concentrated on the highlighted element at the cost of general situational awareness—which could also hinder the exercise of driving autonomy (e.g., [36]). Moreover, studies on safety have long identified user adaptation as a troublesome rebound effect, according to which people get used to the nudges provided by safety systems and end up over-relying on them, increasing risks—and thus, from our perspective, defeating the expected supports to autonomy (e.g., [37]).

In both cases described above, the benefits in terms of human autonomy would entirely depend on full driving automation. Indeed, individuals who are currently excluded from manual driving would also be poor candidates for shared control and supervisory tasks. Similarly, full automation is necessary for AV users to freely engage in other, more fulfilling activities. The tendency for users to become distracted and inadequately supervise the functioning of systems endowed with intermediate levels of automation has tragically led to fatal accidents already. For driving automation to promote autonomy as the freedom to pursue one's own concept of the good life, it must progressively eliminate the need for human intervention and supervision.

5 A Bigger Picture: Transportation Ethics

The ambiguity in the ways in which driving systems mediate various aspects of human autonomy at different levels of automation is clear. Full automation amplifies autonomy in pursuing self-determined life goals while reducing control over operational and tactical decisions. Partial or conditional automation facilitates autonomy over vehicle control but hinders its exercise in strategic terms. The duty to promote and protect the ethical principle of human autonomy can follow paths that are difficult to reconcile. The ambiguity arises from the complexity of the concept of human autonomy and from divergent models of driving automation, which mediate its different aspects differently. This uncertainty leaves open the question of how to incorporate the value of autonomy into driving automation technologies.

The resulting tension raises a very delicate ethical issue. Aligning driving automation with legitimate moral expectations regarding human autonomy is an important task. Violations in this regard risk causing harm, resentment, distrust, and backlash. However, the ambiguous nature of the mediation makes it complicated to move from abstract approvals to practical recommendations for design and policies. What should we do in the face of this ambiguity?

Rather than trying to eliminate it by design or sweeping it under the hype's rug, we must accept it and learn how to manage it. If at the level of the driving system the situation is irredeemably ambiguous, let us broaden the perspective. The amplifications and reductions of autonomy mediated by driving automation, after all, do not by any means exhaust the issue of how the value of human autonomy may be at risk or protected and promoted in our daily interactions with transportation technologies. Instead, they represent only a detail of a much broader and more colourful picture. Putting the discourse on moral autonomy as a crucial value for driving automation into the wider context to which it belongs—that is, into the context of a *transportation ethics*—could prove useful to address the issue discussed in this chapter and better understand what motivates it.

Shifting the perspective from how driving automation is bound to co-shape the exercise and expression of user autonomy to the wider context of transportation systems and their ethical implications allows to identify a further, and critical, layer of human autonomy—that of participated decision-making on social matters pertaining to transportation. Interpreting AVs not in isolation, but as a new ingredient to be wisely added to our transportation recipe calls for *public decision making*—i.e., for "the possibility for, ideally, all people to (indirectly) influence the decision-making process that will determine the design, development, and regulation of CAVs [Connected and Automated Vehicles], and to make sure that their interests and rights are sufficiently reflected in the process" [41, p. 30]. As the public discourse is dominated by hyped narratives presenting AVs as autonomy-enhancing technologies and, more in general, unambiguously beneficial transport innovations, it becomes essential to understand how to foster well-informed, reflective, and autonomous public decision-making on mobility matters.

As long as transportation systems impact on us not only individually, but also as members of society, any new addition to the mix—along with its prospected ethical benefits—requires to be assessed in relation to *social well-being*. To consider what has been discussed so far, issues of accessibility, inclusivity, justice and sustainability clearly appertain to this dimension. Public, participated decision making is critical in this sense to make the best of driving automation—beyond the focus, which dominates the hype, on individual potential customers and their well-being. All the more so, the autonomy of public decision-making as independence from lobbying and commercial interests does stand out as an ethical value that deserves to be protected and fostered—i.e., as yet another aspect of human autonomy to be carefully investigated. This further layer of autonomy, which fully emerges only when the discourse on AVs is placed in the context of transportation ethics, is key to *actually* put this innovative transport technology at the service of ethical values—rather than the other way around.

Competing Interests This chapter is supported by the "Sustainable Mobility Center (Centro Nazionale per la Mobilità Sostenibile—CNMS)", project CN00000023, funded by the European Union NextGenerationEU program within the National Recovery and Resilience Plan (PNRR), Mission 4 Component 2 Investment 1.4 "Potenziamento strutture di ricerca e creazione di 'campioni nazionali di R&S' su alcune Key Enabling Technologies".

References

1. Fossa F (2023) Ethics of driving automation. Artificial agency and human values. Springer, Cham. https://doi.org/10.1007/978-3-031-22982-4
2. Santoni de Sio F (2021) The European Commission report on ethics of connected and automated vehicles and the future of ethics of transportation. Ethics Inf Technol 23:713–726. https://doi.org/10.1007/s10676-021-09609-8
3. Kock B, Lannquist Y (2022) Fulfilling the promise of autonomous vehicles with a new ethics of transportation. In: Jenkins R, Cerný D, Hríbek T (eds) Autonomous vehicle ethics: the trolley problem and beyond. Oxford Academic, New York, pp 390–414. https://doi.org/10.1093/oso/9780197639191.003.0023
4. Marshall W (2024) Killed by a traffic engineer. Shattering the delusion that science underlies our transportation system. Island Press, Washington, DC
5. Ihde D (1979) Technics and praxis. A philosophy of technology. Springer, Cham
6. Ihde D (1990) Technology and the lifeworld. From garden to earth. Indiana University Press, Bloomington

7. Verbeek P-P (2005) What things do. Philosophical reflections on technology, agency, and design. Penn State University Press, University Park
8. Verbeek P-P (2011) Moralizing technology. Understanding and designing the morality of things. The University of Chicago Press, Chicago
9. Rosenberger R, Verbeek P-P (eds) (2015) Postphenomenological investigations. Essays on human technology relations. Lexington Books, London
10. Aagaard J, Friis JKB, Sorenson J, Tafdrup O, Hasse C (eds) (2018) Postphenomenological methodologies. New ways in mediating techno-human relationships. Lexington Books, London
11. Calvo RA, Peters D, Vold K, Ryan RM (2020) Supporting human autonomy in AI systems: a framework for ethical enquiry. In: Burr C, Floridi L (eds) Ethics of digital well-being. Philosophical studies series, vol 140. Springer, Cham. https://doi.org/10.1007/978-3-030-50585-1_2
12. Laitinen A, Sahlgren O (2021) AI systems and respect for human autonomy. Front Artif Intell 4:1–14. https://doi.org/10.3389/frai.2021.705164
13. Formosa P (2021) Robot autonomy vs. human autonomy: social robots, artificial intelligence (AI), and the nature of autonomy. Minds Mach 31:595–616. https://doi.org/10.1007/s11023-021-09579-2
14. Amoroso D, Tamburrini G (2018) The ethical and legal case against autonomy in weapons systems. Glob Jurist 18(1):20170012. https://doi.org/10.1515/gj-2017-0012
15. Amoroso D, Tamburrini G (2021) Toward a normative model of meaningful human control over weapons systems. Ethics Int Aff 35(2):245–272. https://doi.org/10.1017/S0892679421000241
16. Danaher J (2016) The threat of algocracy: reality, resistance and accommodation. Philos Technol 29:245–268. https://doi.org/10.1007/s13347-015-0211-1
17. Danaher J (2018) Toward an ethics of AI assistants: an initial framework. Philos Technol 31:629–653. https://doi.org/10.1007/s13347-018-0317-3
18. Danaher J (2019) The ethics of algorithmic outsourcing in everyday life. In: Yeung K, Lodge M (eds) Algorithmic regulation. Oxford University Press, Oxford, pp 98–117. https://doi.org/10.1093/oso/9780198838494.003.0005
19. Rubel A, Castro C, Pham A (2021) Algorithms and autonomy. The ethics of automated decision systems. Cambridge University Press. https://doi.org/10.1017/9781108895057
20. Bonicalzi S, De Caro M, Giovanola B (2023) Artificial intelligence and autonomy: on the ethical dimension of recommender systems. Topoi 42:819–832. https://doi.org/10.1007/s11245-023-09922-5
21. Christman J (2020) Autonomy in moral and political philosophy. In: Zalta E (ed) The Stanford encyclopedia of philosophy, Fall 2020 edn https://plato.stanford.edu/archives/fall2020/entries/autonomy-moral/
22. Waymo Co-CEO Tekedra Mawakana. Full Interview. Code 2021. https://www.youtube.com/watch?v=xaYCT7woLMo. Last accessed 21 June 2025
23. European Commission: Directorate-General for Research and Innovation (2020) Ethics of connected and automated vehicles: recommendations on road safety, privacy, fairness, explainability and responsibility. Publications Office. https://doi.org/10.2777/035239
24. European Commission: Directorate-General for Communications Networks, Content and Technology (2019) Ethics guidelines for trustworthy AI. Publications Office. https://data.europa.eu/doi/10.2759/346720
25. Nyholm S (2018) The ethics of crashes with self-driving cars: a roadmap II. Philos Compass 13(7):1–10. https://doi.org/10.1111/phc3.12506
26. Nyholm S (2023) Minding the gap(s): different kinds of responsibility gaps related to autonomous vehicles and how to fill them. In: Fossa F, Cheli F (eds) Connected and automated vehicles: integrating engineering and ethics. Studies in applied philosophy, epistemology and rational ethics, vol 67. Springer, Cham. https://doi.org/10.1007/978-3-031-39991-6_1
27. Michon JA (1985) A critical view of driver behavior models: what do we know, what should we do? In: Evans L, Schwing RC (eds) Human behavior and traffic safety. Springer, Boston, pp 485–524. https://doi.org/10.1007/978-1-4613-2173-6_19

28. Moreno C (2015) The 15-Minute City. A solution to saving our time & our planet. Wiley, Hoboken
29. Martens K (2017) Transport justice: designing fair transportation systems. Routledge, New York/London
30. SAE International (2021) J3016_202104—taxonomy and definitions for terms related to driving automation systems for on-road motor vehicles. https://www.sae.org/standards/content/j3016_202104/
31. Schoonmaker J (2016) Proactive privacy for a driverless age. Inf Commun Technol Law 25(2):96–128. https://doi.org/10.1080/13600834.2016.1184456
32. Smids J (2018) The moral case for intelligent speed adaptation. J Appl Philos 35:205–221. https://doi.org/10.1111/japp.12168
33. Contissa G, Lagioia F, Sartor G (2017) The ethical knob: ethically-customisable automated vehicles and the law. Artif Intell Law 25:365–378. https://doi.org/10.1007/s10506-017-9211-z
34. Millar J (2017) Ethics settings for autonomous vehicles. In: Lin P, Abney K, Jenkins R (eds) Robot ethics 2.0: from autonomous cars to artificial intelligence. Oxford University Press, Oxford, pp 20–34. https://doi.org/10.1093/oso/9780190652951.003.0002
35. Santoni De Sio F, van den Hoven J (2018) Meaningful human control over autonomous systems: a philosophical account. Front Robot AI 5(15):1–14. https://doi.org/10.3389/frobt.2018.00015
36. Endsley MR, Jones DG (eds) (2017) Designing for situational awareness: an approach to user-Centered design, 2nd edn. CRC Press, Boca Raton
37. Rudin-Brown C, Jamson F (eds) (2010) Behavioural adaptation and road safety. Theory, evidence and action. CRC Press, Boca Raton
38. Goggin G (2019) Disability, connected cars, and communication. Int J Commun 13:2748–2773. https://ijoc.org/index.php/ijoc/article/view/9021
39. Arfini S, Bellani P, Picardi A, Yan M, Fossa F, Caruso G (2023) Design for Inclusivity in driving automation: theoretical and practical challenges to human-machine interactions and interface design. In: Fossa F, Cheli F (eds) Connected and automated vehicles: integrating engineering and ethics. Studies in applied philosophy, epistemology and rational ethics, vol 67. Springer, Cham. https://doi.org/10.1007/978-3-031-39991-6_4
40. Golbabaei F, Dwyer J, Gomez R, Peterson A, Cocks K, Bubke A, Paz A (2024) Enabling mobility and inclusion: designing accessible autonomous vehicles for people with disabilities. Cities 154:105333. https://doi.org/10.1016/j.cities.2024.105333
41. Santoni de Sio F, Fossa F (2023) Designing driving automation for human autonomy: self-determination, the good life, and social deliberation. In: Fossa F, Cheli F (eds) Connected and automated vehicles. Integrating engineering and ethics. Springer, Cham, pp 19–36. https://doi.org/10.1007/978-3-031-39991-6_2

Fabio Fossa is Assistant Professor (tenure-track) in Moral Philosophy at Politecnico di Milano, Department of Mechanical Engineering. His main research areas are philosophy of technology, robot and AI ethics (with a focus on transportation and social robotics), applied ethics, and the philosophy of Hans Jonas. Among his publications: *Ethics of Driving Automation. Artificial Agency and Human Values* (Springer, 2023).

Personal Avatars and Personal Autonomy

8

Paula Sweeney

Abstract

Personal avatars are avatars that represent actual people. They are trained on a particular person's data to look, behave and interact like them. Such avatars can help overcome problems of capacity and capability—they can help us to be represented in more than one place at a time and they can help us to undertake tasks that we might find difficult or unpleasant. However, as such avatars are not controlled directly by the person that they represent, they also present an attack on our personal autonomy, making us less able to control actions that are undertaken in our name and that, in many cases, may be attributed to us.

Keywords

Personal avatars · Autonomy · Dignity · GenAI · Human representation

In September 2023 Chen Tiru, a Taiwanese influencer who has a following of nearly nine million fans, livestreamed footage of himself eating a Taiwanese delicacy for 15 h straight. His fans were impressed, until it was revealed that it was not Chen Tiru who had performed this feat but his avatar. Chen Tiru had created an avatar version of himself to outsource some of his duties, allowing him to produce and stream more content with the aim of increasing his number of followers [1]. It is not only online influencers who are likely to have personal avatars in the workplace. The company Section School is promoting avatar versions of their teaching staff [2]. These avatars will reportedly take tutorials and will be available to answer student questions 24/7. These are not old-fashioned gaming-style avatars—cartoonish caricatures of an agent—but are incredibly realistic representations of the individual

P. Sweeney (✉)
University of Aberdeen, Aberdeen, Scotland, UK
e-mail: p.sweeney@abdn.ac.uk

M. Anzalone et al. (eds.), *Reconfiguring Human Autonomy*, Integrated Science 40,
https://doi.org/10.1007/978-3-032-14537-6_8

that accurately capture their look, movements, voice and style of communicating. These avatars are so good that it is very difficult, if not impossible, to tell them apart from the agent that they represent.

These new technological entities—I call them *personal avatars*—raise many fascinating philosophical questions. In other recent work I have explored the risk that the actions of our avatars might leave us personally responsible for acts that we did not and would not commit [3]. I have considered whether the use of personal avatars might detract from the value of humans by undermining their uniqueness [4]. Here I am interested in the impact that the use of personal avatars might have on our personal autonomy. In Sect. 1, I explain what a personal avatar is, how it is created, and the areas in which we are beginning to see a rise in such technologies. In Sect. 2, I describe the relation that I understand to exist between an individual and their avatar. I see this as a proxy relation, whereby avatars act as proxies for the agent that they represent. In Sect. 3, I draw on distinctions in the medical humanities literature in order to explicate the impact that the proxy relation can have on personal autonomy. Finally, in Sect. 5, I consider the link between autonomy and dignity and whether the undermining of personal autonomy arising from the use of personal avatars might also result in a loss of dignity.

1 What Is a Personal Avatar?

A personal avatar is an avatar that is intended to represent a particular individual that existed or currently exists. Such avatars are created using GenAI technology that is trained on the data of that individual—data such as text messages, emails, video recordings and so forth. The GenAI technology, which also has core training on language patterns, uses this personal data to create an avatar representation of that particular person—what I call a *personal avatar*.

One form of personal avatar has been fairly widely written about—those variously referred to as ghostbots, deathbots and continuing bond avatars [5]. These are avatars or chatbots that are designed to represent a particular person after they have died. Eugenya Kuyda, now the CEO of the AI company Replika, used the personal data of her friend Roman to create a bot version of him that she and her friends could chat with after he had died suddenly [6]. And James Vlahos spent hours interviewing his father when he was terminally ill so that he could create Dadbot, a chatbot representation of his father [7]. There is much debate over the potential harms and benefits of such avatars with some claiming that they get in the way of a healthy grieving process, and others claiming that they are just a continuation of a longstanding practice of using tools to immortalise the dead, a practice that is generally considered to be an aid to grief [5, 8].

The use of personal avatars is very likely to expand beyond the bereavement market. As Danaher and Nyholm note, viewed as at least partial recreations of ourselves, personal avatars might be seen to make us less scarce [9]. They could be seen to do this in two ways. First, they might be seen to mitigate our temporal and geographical limitations. A person can only be in one place at one time, but an avatar representation can allow a person to be represented in various places at once. For example, a company executive, Robin, might have to fly to Tokyo to meet a new client and yet also be in New York to vote at the company AGM. An avatar could act as Robin's proxy at the AGM while the executive is themselves in Tokyo. Second, an individual person is ontologically unique and, as such, there are circumstances in which they cannot be replaced by another person without loss. An avatar representation of Robin, although not identical with them, can come closest to mimicking their persona and might be accepted as a fitting substitute in some contexts.[1]

There are two distinct needs that personal avatars can seek to fill. The first is the need identified in the bereavement market, a need arising from the fact that sometimes we miss another person so much that we are willing to consider any form of continuing contact with (a representative of) them. We can call these companion-led personal avatars. Companion-led avatars are likely to rise in other areas of loss and a little thought reveals that these go beyond the bereavement market. For example, if one person in a romantic relationship decides that the relationship must end, this can leave the other person in a state close to bereavement. It is certainly conceivable that when a person is in the depths of depression they will turn to avatar technology, wisely or not, as an attempt to alleviate the pain [10].

The second need is very different. It is capacity-led. In some situations, given the limitations of human capacities, it will be useful for a person to create an avatar version of themselves, or to give permission for an avatar version of themselves to be created, for some personal gain. There are examples of such avatars already. As noted above, the online influencer Chen Yiru was exposed as using an avatar version of himself to create content that was falsely branded as livestream footage of him. The demands of his followers were high and he did not have the capacity to create the required content himself, so he outsourced it to his avatar. Sometimes the desire to increase capacity comes, not from the individual themselves, but from some other stakeholder. For example, the SAG-AFTRA Hollywood controversy highlighted plans to use digital replicas of actors in future productions [11]. This was an unpopular proposal. However, the concern seemed to be around ownership of the avatars and the benefit to the person replicated. It is not clear that some actors themselves may not embrace a situation in which an avatar version of them reduces their workload, provided that such an arrangement was appropriately lucrative.

A little consideration will reveal that personal avatars will become a mainstream market. Loneliness is known to be a major societal problem, one that is particularly harmful to the elderly. Loneliness of the elderly is a rising problem because the way that our elders are cared for has changed. In modern society many who have elderly parents are too busy to look after them at home and too busy to spend time with them on a daily basis. As such the burden of care falls on the state. And, although it is recognised as being important, companionship is not the highest priority in a care system that is under pressure, and this leaves some elderly people in need of company. Studies have shown that social robots can help relieve loneliness in care-home

[1] See [4] for a critique of Danaher and Nyholm's framing of the problem of human uniqueness as one of scarcity.

settings [12]. But so too could avatars that are trained on the data of a loved one. You might be too busy at work to spend hours every day chatting or playing scrabble with your elderly parent, but an avatar version of you is never too busy. The avatar trained on your data would be a passable version of you. It is plausible that the avatar version of you would be a preferable companion to a social robot or even a comparatively unknown human carer.

There are obvious benefits in the workplace, too. Consider any job in which personality—a particular style and way of communicating—is a factor relevant to success in the role. Some such roles might be university lecturer, therapist and lawyer. These roles are built upon a knowledge base, but it is also the case that the personality of the role-holder is important to the stakeholder, in this case the student or the client. Performance and success in the role goes beyond a simple conveying of information. It is the *way* that the information is conveyed that makes the person effective in the role and that can be the basis for an important relationship of trust being established. The personality of the person performing the role can make the difference between one who is passable in the role and one who is considered outstanding and is a crucial foundation to the development of a professional relationship. In such professional contexts, where the relationship that an employee builds up with a client is deemed important, we are likely to see the future success of personal avatars in the workplace. It is the importance of a personal relationship that provides a reason for an organisation wanting an avatar version of a *particular employee* rather than a generic off-the-shelf chatbot.

There are three things to note about personal avatars before we continue. The first is that personal avatars as I am understanding them here are different from 'uploads'. An upload is an entity (not currently realisable) that is created when the brain or mind of a person is scanned and uploaded to a server, leading to at least the possibility that the uploaded entity might be identical with the person represented [13]. A personal avatar is different from an upload as it uses GenAI technology and personal data to *represent* the person—the resulting avatar does not purport to *be* identical with the person.[2] The second thing to note is that, unlike virtual reality avatars as they are standardly understood, personal avatars are not directly controlled or teleoperated by a human. Rather they act autonomously, albeit guided by their training data. It is the resulting gap between the training data that determines the proxy action and the intension of the person represented that can lead to responsibility gaps and, as we see below, loss of personal autonomy. Finally, these avatars are importantly different from digital personas. Digital personas are fabricated digital persons. For example, the CGI avatar *Lil Miquela*, a 'virtual fashion icon' who appeared in Time's list of Most Influential People on the Internet, is a fabricated persona who, unlike a personal avatar, is not modelled on the data of one particular

[2] That is not to say that an avatar could not be created using upload technology—it could. That is not the focus of this project. Note that upload technology is, at present, just a theoretical possibility whereas the GenAI technology required for avatar creation is already available. Also, there are personal physical risks involved in using upload technology that do not arise in the use of GenAI technology.

person [14]. In sum, personal avatars are semi-autonomous representations of real people.

2 Avatars as Proxies

As noted above, personal avatars stand in some close relation to the person that they represent. How are we to think of the relation between a personal avatar and the person represented? Some have proposed views that depict avatars as duplicating a person. Danaher and Nyholm, for example, call personal avatars 'digital duplicates' and, in seeing them as potentially capable of reducing the scarcity of an individual human, see the avatar as providing an additional being that could, in principle, be identical with the human [9]. However, such a view is problematic. Although our avatars can stand in very close representation relation to the person they represent, they cannot count as *being* the person. They cannot entirely replace the person without loss of value and the avatars actions are not the actions of the person represented. On the other hand, we may be willing to see the avatar as acting on behalf of the person represented in a way that enables us to count the avatar's actions *as if* they were the actions of that person. Rather than seeing avatars as duplicating humans we should see them as standing in a representation relation to them [4].

But what kind of representation relation is this? I propose it is helpful to consider the basic relation between a person and their avatar as a proxy relation [3]. An avatar is a digital entity that can stand in a proxy relation to a human, making decisions, acting and communicating on their behalf.

Proxy relations are not new and do not rely on technology. In the simplest terms, a proxy is an entity that stands for another entity. However, as Luciano Floridi [15] has pointed out, not everything that 'stands for' is a full proxy. We can distinguish between something that *stands for*, for example a photograph being a representation of you, and something that *stands-in* for, for example, your deputy replacing you at a meeting. A full proxy both stands for and stands in for the person it represents. When it comes to the full proxy relation as it appears with humans, a proxy is traditionally understood as being a person who has the authority to act on behalf of another. For example, you might appoint a proxy to make medical or financial decisions for you. The proposal here is that personal avatars stand in just such a proxy relation to the persons that they represent. They are given authority, either by the agent themselves or by someone else, to represent and act on behalf of the person. In the SAG-AFTRA case, the dispute was over who owned the right to represent the actor. Was it only the actors themselves who could give authority for a replication of them to be created or might this right pass through parts of a contract to a film company? In the case of the continuing bond bot Roman, his friends gave authority for the avatar to represent him, presumably believing that he would not have objected to this if he were alive, while DadBot—made by James Vhalos in partnership with his father—was clearly given the authority to represent by James's father himself.

3 Proxy Relations and Personal Autonomy

Let's return to consider our executive, Robin. Imagine that Robin's employer wants to create an avatar version of her that can represent her in certain contexts. Robin is still valuable to her employer, particularly in her ability to quickly build a rapport with the client but she is highly in demand and does not have time to secure as many clients as her employer would like. Their proposal is that, as the professional relationships that Robin establishes continue, Robin's avatar will supplement this relationship by standing in for her in certain contexts. For while Robin is limited in the amount of time that she can commit to the client, her avatar is not limited in this way. Neither does the avatar require time off or sleep so can, in theory, provide a 24 hours-a-day, seven days-a-week service. Should Robin accept such an arrangement? The answer to this will depend on various factors including how Robin might herself benefit from having an avatar representative. We can imagine that she might be given a pay increase that reflects her willingness to have an avatar represent her and her resulting ability to be more productive, she might have more time to spend on the aspects of the job that are of more interest to her, or she might retain her full salary whilst achieving a greater work-life balance. In any case, one important factor that Robin should consider, a factor that ties to further questions of responsibility, is whether such a human-avatar proxy arrangement will respect her autonomy. When Robin's avatar acts on behalf of her, is her personal autonomy respected?

Autonomy is a philosophically contentious notion. Here, I rely on conceptions of autonomy in the medical setting, as this is an area in which proxies are often used and the impact of this on personal autonomy has been explored. There are at least three relevant senses of autonomy of action [16]. First, one is acting autonomously if one's *actions are free* and not the result of coercion, duress or undue influence. Second, an action can preserve one's autonomy if it is seen as being *authentic*, where authenticity is understood as being consistent with the attitudes, values, dispositions and life plans of the person. We sometimes describe this as the person 'acting in character'. Finally, an action preserves autonomy if it is the result of *effective deliberation*: the person is aware of the alternative courses of action, is aware of the consequences of their choices and chooses the action based on an evaluation. We can say a person has full autonomy when all three of these senses of autonomy are preserved, and partial autonomy is achieved when only one or two senses are preserved.

Does the avatar relation respect personal autonomy? As mentioned above, in considering this question it is useful to draw on distinctions regarding proxy consent in medical contexts. In medical contexts, it is important that practitioners receive consent for medical interventions. But there are situations in which a patient is not able to give that consent directly, perhaps due to reduced mental capacity. Proxy consent can offer a solution, protecting the patient's right to autonomy in securing consent. Bruce L. Miller [17] helpfully lays out a number of ways in which a proxy arrangement can be put in place for consent. Note that Miller uses 'principal' to refer to the person who is represented and 'agent' to refer to the person who is doing the representing.

Specific authorisation occurs when a principal appoints an agent to give or deny permission for a specific procedure at some future time. For example, perhaps a person makes their next of kin their proxy and details exactly what decisions they want them to make in an upcoming medical context. In such cases, the agent can be understood as acting as a direct messenger for the principal—they are simply a conduit for the principal's actions. There is no possibility of a gap between the decision that the principal would have made if they had been able and the decision that the agent communicates on their behalf. As such, this arrangement is one that can preserve the autonomy of the principal in all three senses of free action, authenticity and effective deliberation.

General authorisation with instructions occurs when the principal appoints an agent and gives instructions for them to follow when they are making decisions. Perhaps the principal is going to undergo a range of medical procedures that might lead to various unknown interventions. The principal cannot give specific instructions but they can give more general instructions to guide the agent that they authorise to act on their behalf. In such a case, where the extent and clarity of instructions can vary, the full autonomy of the principal cannot be respected. The action cannot be seen as preserving autonomy as free action, at least not directly, because the action is not directly determined by the principal. However, the action might be considered to preserve autonomy as authenticity if it is consistent with the instructions given and if it clearly reflects the values of the principal. Likewise, if the instructions indicate a clear course of action with regards to the decision in question then effective deliberation could also be met.

General authorisation without instructions occurs when the principal has appointed an agent but has not given any instructions with regards to how they should make decisions. This might occur if a person puts in place a medical proxy before any illness occurs and does not know the nature of the illness that they will be inflicted with. Even in such a case, it is possible to preserve some autonomy if the appointed agent knows the values and attitudes of the principal and acts on those in deciding upon action. Autonomy as authenticity, for example, could still be met in a situation in which no instructions are given if the decision to act does respect the values and attitudes of the principal.

Instructions without authorisation can occur when instructions are left but no agent is appointed and someone has to make a decision. In the medical case this might fall to the next of kin or close friend. In such a case secondary effective deliberation is possible if the instructions are sufficiently detailed and the person making the decision uses the instruction to make the decision. And, again, in such a case, an authentic decision is also possible if the values, attitudes and wishes of the principal are known and respected.

Substitute judgement occurs when no instructions are given and no agent is authorised. In such a case, like above, an agent is appointed and tries to determine what the principal might have wanted. If, again, the agent is a next of kin or close friend who knows the principal well, some aspect of autonomy could be preserved. In such a case the only aspect of autonomy that could perhaps be satisfied is autonomy as authenticity.

Finally, *deputy judgement* occurs when no instructions are given, no agent is authorised and the agent who is appointed does not know the person well enough to determine what they would want to happen. In this case the agent is likely to appeal to what a reasonable person would want in making their decision. As such, none of the aspects of personal autonomy can be respected.

These scenarios give us a good starting point for exploring the various ways that personal avatars as proxies might preserve or undermine an individual's autonomy.

4 Avatar Action and Autonomy

Here we will take each of the medical proxy contexts described above, map it on to an analogous personal avatar proxy context, and consider the likely impact on personal autonomy.

The authorised avatar with explicit orders (Maps to *Specific authorisation*). Above, we considered a situation in which Robin had to meet a client in Tokyo while she was also expected to vote at the company AGM. In such a context we can imagine that Robin knows the agenda of the AGM and can give very detailed instructions to her avatar proxy on how to behave with regards to each point in the proceedings including on the important matter of how to cast a vote on her behalf. All going well, this arrangement can be seen as an instance of *specific authorisation*. Robin has appointed the avatar herself and has given permission for representation regarding a specific event (the AGM). We can understand the avatar as being a direct messenger for Robin and, as such, this arrangement is one in which Robin's autonomy might be preserved in all three senses of free action, authenticity and effective deliberation.[3]

The authorised avatar with access to extensive personal data (Maps to *General authorisation with instructions*). Now, imagine that, having been impressed by the way in which the avatar was able to represent her at the AGM, Robin starts to consider other contexts in which she might deploy her avatar. Again, due to a clash in commitments, Robin cannot attend the monthly executive team meeting. In these meetings the agenda is a bit more flexible so she cannot know exactly what matters will be discussed. However, she is fairly confident in having the avatar represent her as it has been trained on her data which includes her past performance at countless monthly team meetings and various briefing notes relevant to the forthcoming meeting. Robin provides her avatar with some further general instructions regarding the monthly team meeting and authorises it to attend the meeting and act on her behalf. In this case, Robin's full autonomy cannot be preserved because, due to the lack of specific instructions, the avatar will not simply be acting as a conduit for the predetermined decisions of Robin, so it cannot be seen as preserving direct free action. But if the training data is excellent and the instructions clear enough then the

[3] Although even in this situation it is possible that full autonomy could be undermined. See [3] for examples in which, despite the clearest and most detailed possible instructions, an avatar can still be left with having to fill the gaps and make an independent decision.

authenticity of Robin's actions could be preserved and also, assuming the avatar properly applies the instructions in a way that respects Robin's values, then the effective deliberation condition could be met.

The authorised avatar with general guidance (Maps to *General authorisation without instructions*). Now imagine that Robin has been very hands-off when it comes to preparing the avatar for any particular representation context but has given permission for the avatar to represent her. The avatar has been trained on her data, again on her behaviour in a number of meetings and from various briefing notes, but as no specific instructions have been given with regards to action in any particular context then neither effective deliberation nor free action can be preserved. However, as before, if the avatar is able to learn enough about Robin's values and attitudes from their training data, then it is possible that autonomy as authenticity might be preserved.

The unauthorised avatar built with limited personal data, supplemented with testimonials (Maps to *Instructions without authorisation*). This is the kind of case that we are likely to see result in continuing bond avatars. Consider the case of Roman. Roman died without stating any preference regarding having an avatar made of him. Furthermore, although there was an extensive amount of data to use to build the Roman Bot, it is likely that his friends filled in the gaps with some testimonials regarding their view of Roman and his preferences. Again, in this scenario, neither free action nor effective deliberation can be preserved. Perhaps autonomy as authenticity may be preserved if the testimonials are accurate and of a high quality.

The unauthorised avatar with access to extensive data (Maps to *Substitute Judgement*). A different kind of case can come about when a person has not given permission for an avatar to represent them. For simplicity's sake, let's assume that the person has not actively withheld permission for an avatar of them to be created but neither have they given permission. Or perhaps they have given permission but it was coerced. For example, it might be the case that having an avatar becomes an expectation in some work contexts such that refusal could lead to negative consequences for the employee. In such a situation consent is likely to be coerced—a person might be fearful that they will be replaced if they do not consent—and so consent is not given freely. Here, although the avatar could still learn enough about the executive's values to represent them authentically, it seems problematic to describe this as a preservation of the individual's autonomy at all. In such a context, where the proxy relation has not been properly established, we might rather think of the relation as a fraudulent representation, similar to an unauthorised replica or forgery of a painting. In such a scenario, autonomy as freedom is violated, autonomy as effective deliberation cannot be maintained but, again, if the training data is of high quality, it is possible for autonomy as authenticity to be preserved.

The unauthorised avatar with generic guidance (Maps to *Deputy Judgement*). In this kind of case the individual has not given permission and there is very little information available to train the avatar as a faithful representation. Perhaps the organisation puts in place a policy that allows it to make an avatar representative of staff without their permission. They have enough data regarding the way that an employee looks, moves and talks to make an avatar that appears to be the employee,

Table 8.1 Avatars and aspects of autonomy

	Free action	Authenticity	Effective Deliberation
Authorised avatar with explicit orders	YES	YES	YES
Authorised avatar with access to extensive data	NO	POSSIBLE	POSSIBLE
Authorised avatar with general guidance	NO	POSSIBLE	POSSIBLE
Unauthorised avatar with general guidance, supplemented with testimonials	NO	NO	POSSIBLE
Unauthorised avatar with access to extensive data	NO	NO	POSSIBLE
Unauthorised avatar with generic guidance	NO	NO	NO

but they have very little data regarding the kinds of decisions the employee would make, the manner of communicating and their personality. Instead they draw on generic data in making the decision-making aspect of the avatar. The result is an avatar that is recognisably representative of the employee but which respects none of the aspects of personal autonomy (Table 8.1).

In sum, although it is possible for there to be avatar representation that does respect the full autonomy of an individual, many avatars contexts are ones in which full autonomy is not preserved. Furthermore, the usefulness of an avatar is greatly heightened when it is able to operate without explicit orders, making decisions on the basis of either extensive data, general guidance or both. In such cases, as we have seen, the full autonomy of an individual will not be preserved.

There are many consequences of a reduction in autonomy in relation to actions that may be counted as one's own, not least the impact on personal responsibility (See Sweeney 2023a). But here I would like to end by exploring the link between personal autonomy and dignity.

5 Autonomy and Dignity

Autonomy and dignity often seem to go together in public discourse as properties that can be threatened by public policy, but just how they are connected is not always clear. Here I am particularly interested in the following question: if the use of personal avatars can be seen to reduce one's personal autonomy, might it also reduce a person's dignity?

There are (at least) two distinct conceptions of dignity in the philosophical literature. There is the Kantian conception of dignity-as-inner-worth and there is the conception of dignity-as-bearing [18, pp. 6–13]. The Kantian conception of dignity-as-inner-worth identifies dignity as arising from a human's capacity to respect the moral law. On this understanding, dignity and autonomy are, if not synonymous, then very closely connected. The alternative conception of dignity-as-bearing concerns one's ability to gain dignity by behaving in a dignified way. Unlike dignity-as-inner-worth, which is constitutive of persons, dignity-as-bearing can be lost if we find ourselves in humiliating situations.

It is dignity-as-bearing that is likely to be most relevant to our purposes here. Might the loss of autonomy that can arise from the use of a personal avatar lead to humiliation and a resulting loss of dignity-as-bearing? In the first scenario of *authorised avatar with explicit orders*, when a person gives direct orders to an avatar that represents them in a limited context and where full autonomy can be retained, it seems right to say that there is no resultant loss of dignity. The avatar will simply perform as the person themselves would have performed. But in each of the other scenarios there may be a loss of dignity to some degree, and to a high degree in the scenario where no degree of autonomy is retained. In particular, in any scenario in which the avatar is representing an agent, presenting *as being them*, and yet the avatar behaves in a way where the agent's authenticity is not preserved, this opens the agent up to a loss of dignity-in-bearing. This is because, where authenticity is not preserved there is a risk of humiliation. Consider the following example. Suppose an avatar that does not preserve her autonomy represents Robin at a board meeting. Imagine that during that meeting one colleague makes an inappropriate comment about another colleague. If Robin had been attending the meeting herself she would not have laughed at the inappropriate comment and might even have publicly distanced herself from that kind of comment. But the avatar, trained on social niceties of laughing when they detect humour, and not given very detailed training on the workplace interactions of Robin, laughed out loud at the comment. The close representative connection between the avatar and Robin—that it looks, sounds and behaves in many ways like her—has the result that the actions of the avatar reflect on Robin herself. Even if everyone in the meeting is aware that this is an avatar and not in fact Robin, the avatar's actions can still have a negative reflection on Robin. For one thing, the avatar has been trained on Robin's data, so who is to say that the avatar has not simply revealed the true humour of Robin, something that she might herself try to keep hidden in professional contexts. The overall result is that in using a personal avatar as proxy, through a resulting loss of personal autonomy, a person is making themselves vulnerable to a loss of dignity.

6 Concluding Remarks

My purpose was to explore the likely impact of the use of personal avatars on the autonomy of the person that they represent. If personal avatars are understood to stand in a proxy relation to the person that they represent then it seems likely that, in most avatar contexts, the avatar will not be able to preserve the full autonomy of the individual that they represent and in some contexts personal autonomy will be entirely undermined.

Competing Interests The author has no conflicts of interest to declare that are relevant to the content of this chapter.

References

1. Hawkins A (2023) How Chinese influencers use AI digital clones of themselves to pump out content. The Guardian. https://www.theguardian.com/world/2023/nov/06/chinese-influencers-using-ai-digital-clones-of-themselves-to-pump-out-content. Accessed 3 Apr 2025
2. Malmsheimer T (2024) We introduced our first AI avatar professor. Here's why. https://www.sectionschool.com/blog/why-we-use-ai-avatars?utm_source=email&utm_medium=strategybrief&utm_campaign=0413_aiavatar#video. Accessed 3 Apr 2025
3. Sweeney P (2023) Avatars as proxies. Mind Mach 33(3):525–539
4. Sweeney P (2024) Avatars and the value of human uniqueness. Philos Technol 37:118. https://doi.org/10.1007/s13347-024-00811-2
5. Krueger J, Osler L (2022) Communing with the dead online: chatbots, grief, and continuing bonds. J Conscious Stud 29(9–10):222–252
6. Newton C (2016) Speak memory. When her best friend died she rebuilt him using artificial intelligence. The Verge. https://www.theverge.com/a/luka-artificial-intelligence-memorial-roman-mazurenko-bot. Accessed 3 Apr 2025
7. Vlahos J (2017) A son's race to give his dying father artificial immortality. https://www.wired.com/story/a-sons-race-to-give-his-dying-father-artificial-immortality/. Accessed 2 Apr 2025
8. Lindemann NF (2022) The ethics of 'deathbots'. Sci Eng Ethics 28:60. https://doi.org/10.1007/s11948-022-00417-x
9. Danaher J, Nyholm S (2024) Digital duplicates and the scarcity problem: might AI make us less scarce and therefore less valuable? Philos Technol 37:106. https://doi.org/10.1007/s13347-024-00795-z
10. Sweeney P (2023) The ethics of ex-bots. AI Soc 39:3055–3056
11. da Silva J (2024) Actors go on strike over video games AI threat. BBC. https://www.bbc.com/news/articles/c4ngy53qyq8o. Accessed 3 Apr 2025
12. Robinson H, Macdonald B, Kerse N, Broadbent E (2013) The psychological effects of a companion robot: a randomized controlled trial. J Am Med Dir Assoc 14(9):661–667
13. Chalmers D (2014) Uploading: a philosophical analysis. In: Blackford R, Broderick D (eds) Intelligence unbound: the future of uploaded and machine minds. Wiley Blackwell, Chichester/West Sussex, pp 102–117
14. Time Stuff (2018) The 25 most influential people on the internet. https://time.com/5324130/most-influential-internet/. Accessed 3 Apr 2025
15. Floridi L (2015) A proxy culture. Philos Technol 28:487–490
16. Miller BL (1981) Autonomy and the refusal of lifesaving treatment. Hastings Cent Rep 11(4):22–28
17. Miller BL (1982) Autonomy and proxy consent. Ethics Hum Res 4(10):1–8
18. Killmister S (2022) Autonomy and dignity. In: Colburn B (ed) The Routledge handbook of autonomy. Routledge, New York

Paula Sweeney is Senior Lecturer in Philosophy at the University of Aberdeen. Her book *Social Robots: A Fictional Dualism Model* (2023) was published as part of Rowman and Littlefield's series *Philosophy, Technology and Society.* She has published in some of the most highly regarded journals in her field such as *Minds and Machines*, *Ethics and Information Technology* and *Philosophy and Technology.* She is interested in the impact of new technologies on humans. Her most recent work explores the ethics of avatars that represent real people, *personal avatars.*

Part III

Autonomy and AI in Medicine and Health Care. Challenges, Problems, and Opportunities

9 Beyond (Relational) Autonomy: AI Healthcare Through the Lens of Care Ethics

Miranda Boldrini

Abstract

Autonomy is a central concept in both ethics and bioethics, yet the diversity of its theoretical interpretations and practical applications has not been fully integrated into the ethics of artificial intelligence, particularly in healthcare. Despite the emphasis on autonomy in regulatory frameworks and ethical guidelines, the prevailing discourse often lacks conceptual clarity and overlooks distinctions between individualistic and relational accounts of autonomy. This chapter argues that these conceptual differences have significant implications for understanding the ethical challenges posed by AI systems used in clinical decision-making. Drawing from philosophical ethics and bioethics, this chapter proposes a methodological framework based on two axes: the contrast between individualist and relational conceptions of autonomy, and the distinction between informed consent and shared decision-making as models for operationalizing autonomy in medical practice. By applying these lenses to current debates in AI ethics, this chapter seeks to reveal how different normative assumptions lead to different interpretations of ethical risks, responsibilities, and patient agency. Furthermore, it contends that a relational understanding of autonomy can enrich the ethical assessment of AI in healthcare by foregrounding the roles of trust, responsibility, and care.

Keywords

AI algorithms in healthcare · Relational autonomy · Care ethics · Shared decision-making · Trust

M. Boldrini (✉)
Centre François Viète d'épistémologie et d'histoire des sciences et des techniques, Nantes Université, Nantes, France
e-mail: miranda.boldrini@univ-nantes.fr

M. Anzalone et al. (eds.), *Reconfiguring Human Autonomy*, Integrated Science 40,
https://doi.org/10.1007/978-3-032-14537-6_9

1 Introduction

Autonomy is a foundational concept in AI ethics [1], particularly in healthcare applications, where it has a long-standing role as a core principle in medical ethics [2]. The importance of autonomy in this context is underscored by regulatory frameworks and guidelines, such as the European AI Act (2024), which adopts autonomy as a key ethical principle [3].

Despite its widespread use and influence in both policy and ethical discourse, the concept of autonomy has not been systematically examined in AI ethics—especially in relation to healthcare applications. Meanwhile, critical debates within moral philosophy and bioethics have explored various conceptions of autonomy, notably the contrast between individualistic and relational approaches [4]. Yet these discussions have seldom been integrated into the discourse on AI [5, 6].

Drawing from philosophical ethics and bioethics, this chapter critically examines how autonomy is conceptualized and operationalized in the ethics of AI. It explores how different conceptions of autonomy may illuminate distinct ethical challenges posed by AI in healthcare, and how they may point toward different practical responses.

The aim is twofold: first, to consider how relational accounts of autonomy, when integrated into AI ethics, can offer a richer understanding of the ethical implications that AI-based decision-support systems pose for human autonomy in clinical settings; second, to argue that reflecting on these implications through a relational lens opens up further dimensions beyond autonomy —notably trust and responsibility—at both individual and collective levels. Clarifying the underlying assumptions about autonomy may also foster a more reflexive stance within the rapidly expanding field of AI ethics—an area shaped by diverse disciplinary perspectives, and which often lacks systematic conceptual integration and coherence.

To begin, let us consider an example of AI developed for decision-making support in healthcare.

2 A Machine Learning Approach to Identify Patients with Resected Non-Small-Cell Lung Cancer at High Risk of Relapse

A large part of current research on AI for healthcare focuses on the development of machine learning (ML) algorithms for medical prognosis, guided by a personalised approach to medicine. This chapter originates from my experience working within a project of this kind—namely, a project funded by the Era-PerMed programme elaborating "A Machine Learning approach to identify patients with resected non-small-cell lung cancer at high risk of relapse" (MIRACLE). This ML algorithm is being developed with the aim of estimating the risk of relapse in patients with early-stage non-small cell lung cancer (ES-NSCLC) following surgical resection.

In fact, although early-stage NSCLC is potentially curable through surgery, approximately 50% of patients experience disease relapse after the intervention. To prevent relapse, adjuvant therapies may be administered post-surgery. Currently, in oncology, the pathological stage of the tumour is the main prognostic factor used to determine whether adjuvant treatment is appropriate—typically proposed only from stage II onwards. However, pathological staging alone is not a sufficiently accurate predictor. The clinical problem addressed by this project is, therefore, the need to stratify patients using a predictive algorithm, in order to determine more precisely when adjuvant therapy is appropriate for early-stage cancer patients.

The ML model is trained using data derived from the integrated analysis of biological and clinical information, along with radiomic features, from patients with resected early-stage NSCLC, collected in three European countries. The underlying hypothesis is that integrating multi-omics data into an AI model can enhance patient stratification and lead to more accurate prognostic predictions for individuals with NSCLC.

In the development phase of projects such as the one considered here, data-related concerns arise both for patients currently participating in the research and for future patients once the algorithm is implemented. In the AI ethics literature, these concerns are frequently framed in terms of data protection and privacy, justified by the principle of respecting patients' autonomy over their own data, coupled with their right to privacy [1]. A solution consistent with the current literature on the ethics of AI involves, on the one hand, ensuring that data are anonymised and securely stored, and on the other, obtaining informed consent for data use from both current participants and future patients.

Thus, guarantees of privacy and data protection, together with informed consent, are often considered sufficient to uphold the ethical dimension of patients' contribution of data. But can the ethical dimension of patient participation in this type of research be reduced to data protection? And can autonomy be ensured through informed consent?

This is only one example of the kinds of questions that arise in projects of this kind during the development phase. The deployment phase of such algorithms often brings to light additional ethical issues which, as we shall see, are even more frequently framed in terms of patients' autonomy. This kind of ethical questioning in the context of AI for healthcare invites us to interrogate the assumptions underlying the approach to autonomy in this field, and to consider whether additional ethical issues emerge when these technologies are examined through alternative analytical lenses.

3 Autonomy as Self-Determination: The Standard Model of Autonomy and Informed Consent

In ethics and bioethics, following the traditions descending from modern liberal theory (*cf.* [7, 8]), particularly within Kantian and Millian legacies, autonomy is typically conceptualized along two main dimensions. The first dimension concerns the capacities that define an individual—or their choices and actions—as autonomous. These capacities primarily pertain to rationality, i.e., the ability to comprehend facts and situations; the capacity for critical reflection upon one's own preferences and desires insofar as they motivate action; and—according to some theories [9]—the ability to assess their authenticity. The second dimension concerns the external context and aligns with the idea of negative liberty [10, pp. 78–79], namely, the absence of interference within the sphere of individual action and decision-making, whether by other individuals, the state, or other substantive constraints. Thus, from this theoretical framework, the overall picture that emerges is one of autonomy as *self-determination* and *self-governance*, shaped both by *intrinsic characteristics* of the individual (rationality, authenticity) and *extrinsic conditions* (absence of influence or coercion).

This conception of autonomy has supported fundamental political and social advances, particularly as a counterweight to unjustified interference by political authorities or other individuals in the personal sphere of decision and action. In particular, within the Millian tradition [11], autonomy has traditionally been opposed to *paternalism* [12, 13], understood as any intervention in an individual's personal sphere of action and decision that runs contrary to their will and is justified by an appeal to their 'own good', implying that the individual subject to the intervention is deemed incapable of determining the good for themselves.

The critique of paternalism by appeal to a liberal conception of autonomy has been expressed across various domains, particularly in the field of clinical ethics. Indeed, the principle of respect for autonomy has been one of the main pillars of bioethics and clinical ethics since the foundational work of Tom L. Beauchamp and James F. Childress [2]. In this context, patient autonomy of choice is defined in terms of two key elements: the *capacity* for autonomous action (understood as agency), and *independence* from coercion or undue influence (understood as liberty) [2, p. 102]. Respect for autonomy, in this view, is a *prima facie* principle that entails acknowledging the individual's "right to hold views, to make choices, and to take actions based on their values and beliefs" [2, pp. 106–107]. This respect entails both a *negative obligation*—i.e., not to subject autonomous actions to controlling constraints, with exceptions possible in specific cases, such as when others may be harmed—and a *positive obligation*—i.e., to support autonomous decision-making by appropriately disclosing information and fostering comprehension [2, p. 107].

Thus understood, the principle of respect for autonomy in medicine has been primarily operationalized through the procedure of *informed consent* [2, pp. 120 ff.] (see also [14, p. 149]). This procedure reflects an emphasis placed on information. It also highlights the importance attributed to the competencies required for autonomous decision-making, i.e. the ability to understand information and to demonstrate a certain degree of rationality in processing it. Informed consent, as commonly understood in clinical practice, and as theorized by authors such as Beauchamp and Childress, also implicitly conveys a particular view of the patient–physician relationship: a contractual model, in which the asymmetry between physician and patient—an asymmetry that could otherwise give rise to paternalism—is addressed through the formalization of consent as a kind of contract between two equal parties.

4 Relational Autonomy and Shared Decision-Making: Rethinking Patient Agency in Healthcare

The classical liberal conception of autonomy has been the subject of several critiques in philosophy and ethics [15–17]. Critics argue that, by focusing on intrinsic capacities for self-determination and on negative liberty, this standard conception tends to obscure the socially situated and relational dimensions in which individuals are embedded, which fundamentally shape their spheres of decision and action [4, 15, 18]. This critical trajectory spans diverse approaches, including communitarianism, recognition theory, perspectives influenced by Ludwig Wittgenstein's philosophy, and feminist thought. Despite their varied theoretical backgrounds, these perspectives all advocate for a conception of autonomy that acknowledges the social and relational contexts within which individuals act—or are prevented from acting.

Emerging from these critiques is a *relational conception of autonomy*. Although this "umbrella" term [19] refers to a plurality of theories arising from different traditions, it is within feminist philosophy that the relational conception of autonomy has been most systematically developed [4, 20], as an attempt to reclaim and transform a concept that historically served to justify the exclusion of women and minorities from the moral and political domains.

According to Catriona Mackenzie [19]—a feminist philosopher who has significantly contributed to clarifying this concept—the relational conception of autonomy rests on three fundamental assumptions: (a) an anthropology of vulnerability; (b) a relational conception of identity; (c) a question of justice, according to which a just society must provide institutions capable of supporting the development and exercise of autonomy, "particularly for historically oppressed and marginalized social groups" [19, p. 22].

Building on these assumptions, the central ideas that traverse relational theories of autonomy are the following: autonomy is never exercised outside a network of relationships—whether familial or broader social ones—that can either foster or hinder its various expressions; autonomy and dependence are gradual, rather than being derived from intrinsic and absolute characteristics; and individuals do not act and choose as isolated entities, but are situated within relational, social, and historical contexts that are shaped by social factors such as class, race, and gender.

Along with these criticisms of the conception of autonomy, the model of informed consent in medicine—pictured as a contract based on cognitive capacities and information—has also been challenged. Critics argue that it tends to reduce clinical decision-making to a moment of information transfer, detached from the broader context of clinical practices [21, pp. iix–ix]. Moreover, it has been pointed out that being informed and able to choose without explicit coercion does not guarantee the absence of more subtle forms of influence, such as social norms or family expectations [22]. Furthermore, limiting the operationalization of autonomy to the informed consent procedure risks neglecting the broader role that patients and physicians may play in the decision-making process itself—roles that go beyond agreeing with or refusing a proposed treatment, or merely providing information [23, 24].

Relational conceptions of autonomy suggest that informed consent can be reinterpreted as part of an alternative approach to operationalizing patient autonomy in clinical settings, namely that of *shared decision-making* [25, p. 425]. The paradigm of shared decision-making conveys a different picture of the physician–patient relationship: not as a contract, but as a partnership [26]. According to this model, the idea that patients make decisions entirely on their own after receiving the necessary information is, in many cases, considered unrealistic. In practice, patients often make decisions through dialogue with healthcare professionals [27] and within relational contexts that may involve significant others, if the patient so wishes [24, 28]. In this view, such relational elements are not seen merely as potential sources of paternalism, but rather as intrinsic components of the decision-making process itself [23].

While the aim of promoting patient autonomy remains consistent with this model, the relational view requires an expanded focus: from an emphasis on the patient's capacities for giving informed consent—such as being informed, understanding, and deliberating—to a consideration of the physician's responsibilities and abilities, such as offering support, actively listening, and being capable of understanding the patient's values and preferences. Moreover, at the core of this model lies a challenge to the very meaning of consent in clinical practice.

From a care ethics perspective, Joan Tronto [29] distinguishes between two ways of conceiving consent in clinical ethics. The first is the standard idea of informed consent, understood as the realization of patient autonomy. In contrast, Tronto proposes an understanding of consent in clinical practice as the act by which a patient grants authority to healthcare professionals.

This second meaning allows us to consider consent not merely as a procedural act safeguarding autonomy, but as an act that directly entails responsibility for the physician and requires them to be trustworthy. In this view, the act of consent is an expression of relational trust: it requires the physician to act responsibly, in recognition of the authority granted by the patient. Conversely, in the autonomy-based conception of informed consent, once a patient gives consent, the healthcare provider's responsibility is seen as fulfilled—as long as they act in accordance with the consented act.

Thus, in the standard conception, informed consent serves as a mechanism to protect and operationalize patient autonomy through a framework of rights and duties: the patient's right to information, and the physician's duty to provide it and obtain consent. In Tronto's alternative view, however, the moral vocabulary shifts: consent is understood as a grant of authority that intrinsically entails a demand for trustworthiness and responsibility. To consent, from this perspective, is to entrust the physician, and this act of trust generates a specific responsibility: "the act of consent is an acknowledgment of the doctor's responsibility to treat the patient" [29, p. 191]. This second conception does not take the form of a punctual contract establishing symmetrical rights and duties. Instead, it is grounded in a relationship that explicitly acknowledges the asymmetry between patient and physician. Rather than attempting to neutralize this asymmetry, it embraces it within a framework of responsibility and trust.

In this way, understanding consent as a grant of authority allows for the introduction of a dimension of *care* that goes beyond the *cure* aspect of the medical act for which consent is sought. Responsibility and trust, in this context, are regarded as relational practices essential for enabling non-paternalistic forms of care.

5 Autonomy in the Ethics of Medical AI: Information and Consent

As Rafael Calvo and colleagues [30, p. 36] note, "philosophers and technology leaders have gathered and come to consensus over the need to respect and support human autonomy within the design of AI systems [1]. New sets of AI principles codify autonomy-support, mirroring a similar refocus on autonomy within health [2]".

The parallel between AI ethics and clinical ethics is reflected in the strong emphasis placed on information and consent when addressing patient autonomy in this emerging field. This emphasis appears both in discussions of risks and in arguments highlighting the potential benefits of AI. For instance, some authors [31, 32] argue that AI can enhance patient autonomy by increasing the accuracy of predictions and enabling more personalized, preference-based treatments. In this view, greater certainty in predictive information and more individualized care are framed as potential forms of empowerment with respect to patient autonomy.

At the same time, the issue of information is central to concerns about the risks AI poses to autonomy. This is especially evident in debates on data use: informed consent is frequently cited as a key criterion for ethically justifying the use of personal data in machine learning algorithms [30, 33, 34].

Another major issue where informed consent is considered as a tool to protect and promote autonomy concerns whether patients should always be informed that an AI system has been used to support a diagnosis, prognosis, or treatment recommendation [35]. A conception of patients' autonomy based on full disclosure is often adopted: patients should be informed about the risks and benefits of a given AI application and should voluntarily consent to its use [34, p. 861]. This practical approach aligns with the view of autonomy as self-determination exercised through informed consent.

However, this view raises further questions: to what extent should patients be expected to understand how a ML algorithm works in order to give valid consent? Are physicians expected to explain the algorithm's reasoning? Along this line, a common concern is the "black-box" nature of many ML models—an opacity that is widely regarded as a fundamental barrier to fully informed consent, and therefore a threat to both patient and physician autonomy [36, 37]. This concern has led to a strong emphasis, in both the scientific literature and policy guidelines, on transparency and explainability as essential conditions for promoting human autonomy via informed consent [38]. In some cases, this lack of explainability has even been cited as a reason to limit the use of AI, or to determine which classes of algorithms should be approved on a case-by-case basis [34], given the challenges they pose for both patients and physicians.

Yet some authors have raised doubts about the focus on transparency of this debate, arguing that it is based on a problematic assumption about what counts as relevant information for physicians and patients. Thomas Grote [39], for example, suggests that the primary obstacle to informed consent is not algorithmic opacity itself, but rather clinical communication. As he argues, "even if the internal processes of the ML model are opaque, patients can be provided with a detailed higher-order account about the model's functionality" [39, p. 16]. This could include, for instance, information about the types of data used (e.g., medical records, data sources, etc.), the prediction target (e.g., risk of relapse), the metrics used to measure the algorithm's performance, and whether results vary across different subgroups. According to Grote, this kind of element can enable physicians to communicate effectively with patients, allowing them to assess the relevance and reliability of the algorithm's output—even if the algorithm functions as a black box.

This argument has the merit of challenging some of the simplistic assumptions that underlie general claims about opacity as an irreducible limit and explainability as the universal solution for deploying AI in healthcare and other fields. However, it may still be vulnerable to some of the same criticisms directed at the explainability paradigm. In practice, are physicians and patients truly able to access, understand, and meaningfully engage with the statistical features of these models? Can clinical communication realistically integrate this level of complexity? Or, more constructively, we might ask: how could healthcare institutions be structured, and professionals supported, to make this kind of meaningful, autonomy-supportive communication possible?

This brief overview of a key debate surrounding autonomy and AI in healthcare reveals a defining feature: the central role of information in shaping conceptions of patient autonomy, as operationalized through consent. Yet some critics have expressed concerns about what has been identified as an "informational model" of autonomy in AI ethics [5], warning that such a model may be overly narrow. Does focusing on information alone adequately capture the various ways in which patient autonomy may be undermined—or enhanced—in the context of AI-supported decision-making?

6 Relational Autonomy and the Ethics of Medical AI

Along these lines of criticism, a few authors have begun to explore how adopting a relational perspective on autonomy might reshape the way ethical issues in AI healthcare are addressed. To date, two main lines of inquiry can be identified. The first [5, 6] focuses on the socially situated use of technologies such as remote health monitoring and recommendation systems, and their impact on users' autonomy. The second [25] seeks to enrich the informed consent model by embedding it within a more comprehensive framework of shared decision-making, in order to better account for the dynamics of the therapeutic relationship when AI is introduced. In what follows, I examine these two directions in greater detail, then propose a complementary perspective, following Tronto's ethics of care, which reconsiders the question of autonomy within a broader framework of trust and responsibility.

6.1 Relational Autonomy and AI Health Monitoring

One of the most comprehensive attempts to adopt the relational conception of autonomy in this field is developed by American bioethicist Anita Ho [6], who focuses on AI-driven health monitoring tools. Ho argues that promoting autonomy in the context of such technologies requires more than simply obtaining informed consent. From a relational perspective, it is essential to consider how forms of influence that are not explicitly visible—such as social domination, structural injustice, exploitative environments—can undermine individuals' autonomy, understood as the possibility to develop and live according to one's own values and identity.

For example, while healthcare systems and commercial developers increasingly promote remote health monitoring technologies as efficient and precise tools for enhancing patients' autonomy—by, for instance, promoting independent living for the elderly and persons with disabilities—these same technologies can exert new forms of control over individuals' bodily experience and constrain their choices. This occurs, for instance, when they promote one way of living as more valuable than others, regardless of their values or practical identity.

According to Ho, an adequate conception of autonomy in this context must take into account how individuals' decisions regarding AI health monitoring tools are entangled in a complex network of social relationships, professional practices, societal norms, policies, and regulations. These factors not only shape identity but also influence individuals' concrete ability to act autonomously when making healthcare decisions involving technologies.

Therefore, evaluating autonomy in this context requires a relational lens that considers not only individual conditions—such as cognitive capacities or freedom from coercion—but also the broader structural conditions that enable or hinder people's ability to form and act upon their own conception of the good life. In Ho's view, starting from human vulnerability, dependence, and social needs—whether temporary or constant—the idea of relational autonomy highlights the importance of building trustworthy governance structures that recognize individuals' needs for various forms of support throughout one's life. In this framework, the analysis of autonomy shifts focus: from the evaluation of internal rational capacities and access to information, to the broader social, institutional, and practical conditions that make autonomous deliberation possible.

Another example of a relational approach in the field of ethics of AI for healthcare is provided by Simona Tiribelli [5, 40]. Like Ho, Tiribelli focuses on the exponential growth of health recommender systems (HRSs) in the mobile health (m-health) industry, particularly in relation to the promotion of healthy, active ageing. According to Tiribelli, autonomy is a central ethical value for ageing well, but its current interpretation within HRS design is often reduced to "informational self-determination".

This narrow view of autonomy emphasizes privacy, consent, and the explainability of algorithmic recommendations, while overlooking the broader social and motivational conditions that enable individuals—especially older adults—to act autonomously. In response, Tiribelli proposes a redefinition of the principle of autonomy from a relational perspective. Her account seeks to integrate the notion of authenticity—intended as making decisions that genuinely reflect one's identity and values—with a socio-relational dimension of self-respect and self-trust grounded in social recognition. Indeed, for Tiribelli, self-respect and self-trust, as socio-relationally grounded phenomena, are not merely desirable, but essential for maintaining agency, particularly in later stages of life. From this standpoint, fostering autonomy requires more than protecting individual choices: it involves designing systems that support meaningful relationships, acknowledge interdependence, and create conditions in which individuals can engage reflectively with the technologies shaping their health-related decisions. In this view, relational autonomy serves both as a theoretical lens and a normative tool for rethinking how AI systems can truly support human flourishing in ageing societies.

Both Ho and Tiribelli show that a relational conception of autonomy helps to highlight the social and relational contexts that can either support or undermine the autonomy of patients using AI-driven healthcare tools—whether for health monitoring or recommendation purposes—moving beyond a conception of autonomy limited to rational capacities and information. However, even within their frameworks, autonomy still seems to be conceptualized as the capacity to make choices individually: an internal or private process that can be shaped or influenced from the outside. What appears to be missing, though, is a relational and shared dimension of the decision-making process itself, in healthcare as in other contexts, and a consideration of how this dimension is affected by the introduction of AI tools. In other words, these accounts tend to portray choice as a private, individual act situated within a relational environment that may exert either a supportive or paternalistic influence. Yet, the relational network itself is not fully integrated into the structure of the choice process.

6.2 Shared Decision-Making and AI

A complementary perspective may be found in another approach that adopts relational autonomy in this field, offered by Giorgia Lorenzini and colleagues [25], who focus on the clinical decision-making process itself—particularly its relational structure—and propose a reconceptualization of autonomy and AI through the lens of *shared decision-making* (see also [41–44]). Lorenzini and colleagues identify two essential elements for shared decision-making: effective communication and respect for voluntary choice. How does the introduction of AI algorithms into the clinical relationship affect these two elements? According to these authors, this impact unfolds on two levels: (a) the interaction between physicians and AI, and (b) the communication between physicians and patients regarding the use of AI.

The first level—interaction between physicians and AI—raises the issue of *professional autonomy*. According to Lorenzini and colleagues, professional autonomy requires both clinical and technological competence, including the ability to assess AI-generated recommendations and detect potential errors. This highlights the need to define what level of understanding is necessary for physicians to integrate AI outputs into their practice while maintaining critical judgment.

It is precisely at this point that the problem of the 'black box' becomes salient: the opacity of some AI systems may hinder the physician's ability to understand how, and on what basis, the AI produces a given output. As a result, the demand for explainability must be addressed realistically—that is, in a way that accounts for the variability in physicians' understanding and technological proficiency, and how these can be supported and improved. Such variability may depend on age, context, social background, and individual familiarity with digital technologies, that should be taken into account.

The second level concerns the physician–patient relationship. Lorenzini and colleagues propose a reconceptualization of shared decision-making in light of AI integration as a shift from the traditional *dyadic model* (physician–patient) to a *triadic model* involving physician, patient, and AI system. Within this triad, an ethical and epistemic risk emerges: that AI might become overly dominant in clinical decision-making. According to the authors, this risk must be counterbalanced by physicians, who should remain attentive to the patient's values and preferences—elements that may not be fully captured by the algorithm—and avoid phenomena such as overreliance on AI-generated outputs.

A first question raised by this account concerns how such physician attentiveness can be ethically justified and motivating. Moreover, the introduction of AI into the physician–patient dyad can be problematized with reference to two additional aspects. First, shared decision-making, when grounded in a relational conception of autonomy that aims to be realistic and pragmatic rather than merely ideal, must also take into account the patient's relational network, which often enters into the deliberative process in healthcare [23, 24, 27]. Second, although the authors highlight professional autonomy and dialogue as key components of shared decision-making, they appear to overlook another constitutive element of this model: *trust*. Introducing the question of trust can shed light on how to make sense of, and ethically justify, the physician's role as a counterbalancing figure and as guarantor of the patient' s preferences within the triadic model.

7 Beyond Autonomy, Towards Trust (and Care)?

Considering trust in this context allows to introduce two further dimensions. First, the integration of AI decision-support tools may risk undermining patients' *trust in their physicians*—yet, trust is a necessary condition for shared decision-making. Second, there is the issue of *trust in the algorithm itself*. According to several theoretical approaches, attributing trustworthiness to an algorithm may involve a category mistake: trust presupposes an agent's intentionality and responsibility—qualities that AI systems do not possess.

In the field of AI ethics and regulatory frameworks, however, "trust" is another concept increasingly presented as a key criterion for evaluating the ethical acceptability of machine learning algorithms [45]. But what does it mean to *trust* an algorithm in this context? Often, the focus is placed on technical dimensions such as the quality of training data or the opacity of "black box" systems. Within this framework, the lack of transparency and explainability is viewed as a barrier to trust, thereby justifying efforts to enhance algorithmic intelligibility [38].

As discussed above, the issue of trust becomes particularly salient in healthcare, where it intersects with a longstanding ethical concern: the trust patients place in physicians, and the moral and epistemic trustworthiness of medical professionals [46]. This form of trust is foundational to the therapeutic relationship, shaping both clinical decisions and outcomes. But can an algorithm truly be considered trustworthy in this same ethical sense?

Some authors [36, 47, 48] argue that the concept of trust is inapplicable to AI systems. Philosophically, trust is not reducible to mere reliability or functional accuracy. For instance, according to Annette Baier's account [49], trust involves a particular kind of dependence on another's goodwill: "When I trust another, I depend on her goodwill toward me", and "where one depends on another's goodwill, one is necessarily vulnerable to the limits of that goodwill" [49, pp. 234–235]. Trust, on this view, implies a vulnerability to harm if the trusted party fails to meet the responsibilities inherent in trust [50]. Trust and responsibility are intrinsically connected in this framework. Based on this understanding, AI systems—lacking intentionality, moral agency, and the capacity to recognize and respond to human vulnerability—can be *reliable*, but not *trustworthy* [36, 47, 48, 51].

On the one hand, this conception of trust challenges the idea that AI tools can, *per se*, be a locus of ethical trust. On the other hand, it helps make sense of Lorenzini and colleagues' claim that physicians should counterbalance the role of AI in the triadic structure of medical decision-making. Following Tronto's conception of consent in medicine [29], this emerges not merely as a professional duty, but as a form of ethical responsibility assumed when physicians ask for the patient's trust as they seek consent on the basis of their epistemic authority. From a relational perspective, trust also extends beyond the individual interaction: physicians act within a broader social and institutional framework, and may thus be considered responsible for mediating the role of AI within shared decision-making.

This perspective aligns with Tronto's ethics of care, in which responsibility and trust are not confined to personal virtue or professional conduct but are embedded in a broader conception of the individual as socially situated and context-dependent. Her framework allows us to broaden the ethical lens of trust to include the political and institutional conditions under which healthcare decisions are made. Indeed, what matters for trust in this context is not only the technical reliability of the algorithm, but the entire ecosystem in which it is designed and implemented—that is, how systems of responsibility, transparency, and oversight are structured to protect and support those who depend on these technologies [51].

This shift from individual to institutional trust resonates with relational conceptions of both autonomy and trust [50]. It allows us to expand the ethical inquiry on AI beyond technical criteria such as accuracy, explainability, and transparency. Moreover, rather than focusing solely on the informational dimension of autonomy or the functional reliability of AI, a relational approach to autonomy also allows to consider the social and institutional contexts in which such technologies are embedded in networks of trust and responsibilities—and how these contexts shape the ways in which AI is used and understood, both collectively and individually.

Competing Interests This study was funded by the Era-PerMed MIRACLE Project (Grant No. ERP-2021-23680708).

References

1. Floridi L, Cowls J, Beltrametti M et al (2018) AI4People—an ethical framework for a good AI society: opportunities, risks, principles, and recommendations. Minds Mach 28:689–707. https://doi.org/10.1007/s11023-018-9482-5
2. Beauchamp TL, Childress JF (2013) Principles of biomedical ethics, 7th edn. Oxford University Press, New York
3. Frischhut M, Werner-Felmayer G (2024) Medical ethical principles against the background of 'European values'. Medicine (Baltimore) 52:409–412. https://doi.org/10.1016/j.mpmed.2024.04.012
4. Mackenzie C, Stoljar N (2000) Relational Autonomy: Feminist perspectives on autonomy, agency, and the social self. Oxford University Press, New York
5. Tiribelli S (2023) The AI ethics principle of autonomy in health recommender systems. Argumenta 16:1–8
6. Ho A (2023) Live like nobody is watching: relational autonomy in the age of artificial intelligence health monitoring, 1st ed. Oxford University Press, New York
7. Dworkin G (2008) The theory and practice of autonomy. Transferred to digital print. Cambridge University Press, Cambridge
8. Feinberg J (1989) Autonomy. In: Christman JP (ed) The inner citadel: essays on individual autonomy. Oxford University Press, New York, pp 27–53
9. Frankfurt HG (1971) Freedom of the will and the concept of a person. J Philos 68:5–20
10. Jennings B (2009) Public health and liberty: beyond the Millian paradigm. Public Health Ethics 2:123–134
11. Mill JS (1859) On liberty. Broadview Press
12. Dworkin G (1972) Paternalism. Monist 56:64–84
13. Coons C, Weber M (2013) Paternalism: theory and practice. Cambridge University Press
14. O'Neill O (2007) Autonomy and trust in bioethics. Transferred to digital printing. Cambridge University Press, Cambridge
15. Christman J, Anderson J (2005) Autonomy and the challenges to liberalism: new essays. Cambridge University Press, Cambridge
16. Childress JF, Quante M (2021) Thick (concepts of) autonomy: personal autonomy in ethics and bioethics. Springer International Publishing
17. Jouan M, Laugier S (2009) Comment penser l'autonomie ? entre compétences et dépendances. Presses universitaires de France, Paris
18. Taylor C (1991) The ethics of authenticity. Harvard University Press

19. Mackenzie C (2014) Three dimensions of autonomy: a relational analysis. In: Veltman A, Piper M (eds) Autonomy, oppression, and gender. Oxford University Press, New York, pp 15–41
20. Oshana MAL (1998) Personal autonomy and society. J Soc Philos 29:81–102
21. Manson NC, O'Neill O (2007) Rethinking informed consent in bioethics. Cambridge University Press
22. Ho A (2008) The Individualist Model of Autonomy and the Challenge of Disability. Journal of Bioethical Inquiry 5:193–207. https://doi.org/10.1007/s11673-007-9075-0
23. Dove ES, Kelly SE, Lucivero F et al (2017) Beyond individualism: is there a place for relational autonomy in clinical practice and research? Clin Ethics 12:150–165. https://doi.org/10.1177/1477750917704156
24. Ells C, Hunt MR, Chambers-Evans J (2011) Relational autonomy as an essential component of patient-centered care. Int J Fem Approaches Bioeth 4:79–101. https://doi.org/10.3138/ijfab.4.2.79
25. Lorenzini G, Arbelaez Ossa L, Shaw DM, Elger BS (2023) Artificial intelligence and the doctor–patient relationship expanding the paradigm of shared decision making. Bioethics 37:424–429. https://doi.org/10.1111/bioe.13158
26. Godolphin W (2009) Shared decision-making. Healthc Q 12:e186–e190. https://doi.org/10.12927/hcq.2009.20947
27. Gómez-Vírseda C, De Maeseneer Y, Gastmans C (2019) Relational autonomy: what does it mean and how is it used in end-of-life care? A systematic review of argument-based ethics literature. BMC Med Ethics 20:76. https://doi.org/10.1186/s12910-019-0417-3
28. Prainsack B (2017) Personalized medicine: empowered patients in the 21st century ? New York University Press, New York
29. Tronto JC (2008) Consent as a grant of authority: a care ethics reading of informed consent. In: Lindemann H, Verkerk M, Urban Walker M (eds) Naturalized bioethics: toward responsible knowing and practice. Cambridge University Press
30. Calvo RA, Peters D, Vold K, Ryan RM (2020) Supporting human autonomy in AI systems: a framework for ethical enquiry. In: Burr C, Floridi L (eds) Ethics of digital well-being. Springer International Publishing, Cham, pp 31–54
31. Earp BD, Porsdam Mann S, Allen J et al (2024) A personalized patient preference predictor for substituted judgments in healthcare: technically feasible and ethically desirable. Am J Bioeth 24:13–26. https://doi.org/10.1080/15265161.2023.2296402
32. Briganti G, Le Moine O (2020) Artificial intelligence in medicine: today and tomorrow. Front Med 7:509744. https://doi.org/10.3389/fmed.2020.00027
33. Jobin A, Ienca M, Vayena E (2019) The global landscape of AI ethics guidelines. Nat Mach Intell 1:389–399. https://doi.org/10.1038/s42256-019-0088-2
34. Liao SM (2023) Ethics of AI and health care: towards a substantive human rights framework. Topoi 42:857–866. https://doi.org/10.1007/s11245-023-09911-8
35. Lorenzini G, Shaw DM, Arbelaez Ossa L, Elger BS (2023) Machine learning applications in healthcare and the role of informed consent: ethical and practical considerations. Clin Ethics 18:451–456. https://doi.org/10.1177/14777509221094476
36. Hatherley JJ (2020) Limits of trust in medical AI. J Med Ethics 46:478–481. https://doi.org/10.1136/medethics-2019-105935
37. Rubel A, Castro C, Pham A (2021) Algorithms and autonomy: the ethics of automated decision systems. Cambridge University Press
38. Adeniran AA, Onebunne AP, William P (2024) Explainable AI (XAI) in healthcare: Enhancing trust and transparency in critical decision-making. World J Adv Res Rev 23:2447–2658. https://doi.org/10.30574/wjarr.2024.23.3.2936
39. Grote T (2024) Machine learning in healthcare and the methodological priority of epistemology over ethics. Inquiry 68(4):1218–1247. https://doi.org/10.1080/0020174X.2024.2312207
40. Tiribelli S, Calvaresi D (2024) Rethinking health recommender systems for active aging: an autonomy-based ethical analysis. Sci Eng Ethics 30:22. https://doi.org/10.1007/s11948-024-00479-z

41. Abbasgholizadeh Rahimi S, Cwintal M, Huang Y et al (2022) Application of artificial intelligence in shared decision making: scoping review. JMIR Med Inform 10:e36199. https://doi.org/10.2196/36199
42. Hassan N, Slight R, Bimpong K et al (2024) Systematic review to understand users perspectives on AI-enabled decision aids to inform shared decision making. NPJ Digit Med 7:332. https://doi.org/10.1038/s41746-024-01326-y
43. McDougall RJ (2019) Computer knows best? The need for value-flexibility in medical AI. J Med Ethics 45:156–160. https://doi.org/10.1136/medethics-2018-105118
44. Bjerring JC, Busch J (2021) Artificial intelligence and patient-centered decision-making. Philos Technol 34:349–371. https://doi.org/10.1007/s13347-019-00391-6
45. Zuchowski LC, Zuchowski ML, Nagel E (2024) A trust based framework for the envelopment of medical AI. NPJ Digit Med 7:230. https://doi.org/10.1038/s41746-024-01224-3
46. Grote T, Berens P (2020) On the ethics of algorithmic decision-making in healthcare. J Med Ethics 46:205–211. https://doi.org/10.1136/medethics-2019-105586
47. Budnik C (2025) Can we trust artificial intelligence? Philos Technol 38:10. https://doi.org/10.1007/s13347-024-00820-1
48. Nurock V (2023) Confiance et éthique. In: Citton Y, Lechner M, Masure A (eds), Angles morts du numérique ubiquitaire. Presses universitaires de Paris Nanterre, Nanterre, 79–94. https://doi.org/10.4000/11tvx
49. Baier A (1986) Trust and antitrust. Ethics 96:231–260
50. Baghramian M, Petherbridge D, Stout R (2020) Vulnerability and trust: an introduction. Int J Philos Stud 28:575–582. https://doi.org/10.1080/09672559.2020.1855814
51. Kerasidou CX, Kerasidou A, Buscher M, Wilkinson S (2022) Before and beyond trust: reliance in medical AI. J Med Ethics 48:852–856. https://doi.org/10.1136/medethics-2020-107095

Miranda Boldrini is currently a postdoctoral researcher in the ethics of AI in healthcare as part of the PRIMUS project at Nantes University. Her research explores the concepts of autonomy and vulnerability within the fields of AI ethics in healthcare, care ethics, disability studies, and Wittgensteinian ethics. Her recent publications include the book *Iris Murdoch: Une épistémologie morale réaliste* (Vrin, 2025) and the co-edited volume *L'autonomie à l'épreuve du handicap, le handicap à l'épreuve de l'autonomie* (Presses Universitaires Saint-Louis Bruxelles, 2025).

Balancing Autonomies: The Impact of AI on Patient and Physician Decision-Making in Healthcare

10

Andrea Berber and Jelena Mijić

Abstract

This chapter explores how integrating artificial intelligence into clinical settings impacts the autonomy of physicians and patients. Focusing on advanced systems such as large language models and clinical decision support systems, this chapter examines the dual impact of AI on clinical decision-making. For physicians, computational and value opacity, regulatory ambiguity, automation bias, and deskilling can compromise professional independence. Simultaneously, transitive paternalism and relational opacity, which hinder truly informed consent and participatory decision-making, challenge patient autonomy. Nevertheless, artificial intelligence also offers opportunities to enhance autonomy: for physicians, through administrative relief as well as diagnostic assistance, for patients, via wearables, chatbots, and AI surrogates that promote personalized care and independent health management. This chapter emphasizes that safeguarding autonomy in AI-driven medicine requires more than transparency; it demands shared ethical responsibility among developers, institutions, and clinicians. Autonomy must be understood beyond informed consent, focusing on explicability, meaningful engagement, and alignment with patient values and well-being. Ultimately, AI should serve patient well-being.

Keywords

Artificial intelligence · Clinical decision-making · Patient autonomy · Physician autonomy · Medical ethics

A. Berber (✉) · J. Mijić
University of Belgrade, Belgrade, Serbia
e-mail: andrea.berber@f.bg.ac.rs

M. Anzalone et al. (eds.), *Reconfiguring Human Autonomy*, Integrated Science 40,
https://doi.org/10.1007/978-3-032-14537-6_10

1 AI and Clinical Decision-Making

Clinical decision-making encompasses a broad range of cognitive processes in patient care, including diagnosis, treatment planning, and prognosis assessment. Traditionally, the integration of artificial intelligence (AI) in this domain has been most visible through clinical decision support systems (CDSSs), which assist clinicians by suggesting diagnoses, predicting outcomes, recommending treatments, and aiding in tasks such as triage and patient prioritization [1–3]. However, AI's impact on clinical decision-making extends beyond traditional CDSSs, with a broader array of tools including predictive processing programs, patient-facing health applications, and large language models (LLMs). These tools offer multifaceted support to clinicians and empower patients to navigate the complexities of clinical decision-making, influencing medical decisions from both sides of the healthcare relationship.[1]

Crucially, the application of AI in clinical decision-making is not merely a reflection of the broader proliferation of AI but rather a targeted response to pressing issues such as diagnostic errors.[2] Given the complexity of clinical reasoning, it is unsurprising that efforts to develop computational tools to assist clinicians date back several decades. In particular, the development of CDSSs marked a pivotal point in the systematic enhancement of clinical judgment. Early rule-based CDSSs (1970s) offered valuable insights, but limitations in adaptability and integration hindered their broader adoption. In contrast, today's AI-CDSSs leverage machine learning (ML) and LLMs, allowing for more flexible, data-driven support in clinical decision-making [4, 10, 11].

These modern systems offer dynamic adaptation and the capacity to process vast amounts of patient data and medical literature. For example, ML algorithms are used for the automated analysis of medical images (e.g., detecting tumors in X-rays or MRI scans) and for predicting a patient's risk of developing certain conditions (e.g., cardiovascular disease) [12]. On the other hand, LLMs increasingly serve for applications such as generating patient summaries, assisting in literature reviews, and suggesting diagnosis and treatment options. Notably, in February 2025, it was reported that the LLM DeepSeek was already being used in 90 tertiary hospitals in

[1] The potential of Natural Language Processing (NLP)—and more specifically, LLMs, as pre-trained AI systems designed for flexible adaptation across domains and tasks [4]—is recognized for supporting the further development of health chat tools, such as medical chatbots, that can provide clinical feedback. LLMs, such as ChatGPT, are increasingly recognized as an effective source of medical information for the public, demonstrating the ability to provide high-quality and accessible answers. Research indicates that LLMs can successfully identify patients' needs, accurately answer medical questions, and even be more reliable than standard search engines such as Google in certain areas of medical knowledge [5, 6]. While offering valuable opportunities for clinical feedback, LLMs currently suffer from critical limitations, including hallucinations and sycophantic tendencies [7], which compromise their reliability in high-stakes medical contexts.

[2] The diagnostic process refers more specifically to the sequential activities aimed at identifying a patient's health problem through information gathering, integration, and interpretation—based on clinical history, physical examination, diagnostic testing, and consultations with other healthcare professionals [8, 9].

China, supporting medical imaging analysis, clinical decision-making, and hospital administration [13].

Against this backdrop, AI is increasingly regarded as a means to enhance clinical decision-making, offering novel ways to integrate and interpret clinical information and, ultimately, support physicians in formulating more accurate and timely diagnoses and treatment choices. These advancements also promise to better inform patients about their health conditions and available options, enabling more active participation in shared decision-making. However, they also prompt critical reflection on the changing dynamics of clinical decision-making, particularly concerning the autonomy of both physicians and patients in an increasingly AI-supported healthcare environment.

2 The Concept of Autonomy in Medicine

The word "autonomy" comes from the Greek words *auto*, meaning "self," and *nomos*, meaning "law." It was first used to describe Greek city-states whose citizens made their own laws, free from external rule. This political concept evolved into a broader concept of personal autonomy, signifying self-determination through one's own reasoning and volition. In philosophical theorizing, the term "autonomy" lacks a clear meaning and is interpreted in various ways. This ambiguity often makes it unclear whether discussions concern different facets of a single concept or distinct concepts altogether. In order to provide a focused analysis of AI's impact, this work will treat autonomy as an ethical principle.

In contemporary bioethics, autonomy, or respect for autonomy, has become one of the four foundational principles—alongside beneficence, nonmaleficence, and justice—guiding ethical decision-making in clinical practice, research, and health policy. While nonmaleficence and beneficence historically dominated traditional medical ethics, autonomy and justice have gained prominence only in recent decades [14]. The advancements in modern bioethics, shaped by the emergence of individual rights and reactions to medical abuses during and after World War II, brought autonomy to the forefront of ethical discourse [15].

In the context of the Fourth Industrial Revolution, characterized by rapid advancements in AI, bioethics faces renewed challenges and opportunities. Luciano Floridi and Josh Cowls [16] argue that AI's ethical governance requires *explicability*—namely, transparency in system function and accountability for its outputs. This principle, they note, complements traditional bioethics by clarifying AI's impact on human autonomy and assigning responsibility for errors. These technological shifts have sparked new debates about human rights in the digital age [17] and prompted a critical reevaluation of autonomy in clinical settings for both patients and physicians.

In the medical context, autonomy typically refers to patients' autonomy and is most commonly expressed through *informed consent*. Unlike the paternalistic physician-patient relationship, which is grounded in medical authority, the modern model emphasizes a patient-centered approach, in which the patient retains final authority over treatment decisions. Informed consent, therefore, embodies the shift toward respecting patient autonomy and aligning care with individual values and beliefs [18]. Within this framework, physicians must ensure that competent patients receive adequate information about the diagnosis, possible treatments, associated risks and benefits, and available alternatives. Also, the patients need to understand this information, which supports a process of voluntary and informed decision-making [19]. Informed consent is both an ethical obligation and a legal requirement in many jurisdictions, where its absence may result in legal consequences [20].

Following a commitment to autonomy, when a patient's decision-making capacity is significantly diminished, as often occurs in end-of-life care, treatment decisions should be guided by an *advance directive*, if one exists. Such directives typically document the patient's treatment preferences or designate a *surrogate decision-maker* expressed in anticipation of future incapacity [18, 21].

However, autonomy in medicine extends beyond the patients; it also critically applies to the physician's role. Physician autonomy refers to the ability of doctors to make independent decisions based on their clinical expertise and moral conscience. Edmund D. Pellegrino [22] defines this autonomy through two dimensions: personal autonomy, the right to act in line with one's conscience, and professional autonomy, the freedom to apply medical judgment without undue interference. Florian Funer and Urban Wiesing [23] stress that professional autonomy is vital to ethical care, enabling physicians to uphold core medical values like beneficence and respect for human dignity, even in the face of institutional or economic pressures. Without this autonomy, physicians risk becoming system-bound agents who are unable to put priority on patient needs.

There is a natural tension between patient and physician autonomy [22]. While modern ethics rightly emphasizes patient self-determination in response to past paternalism, this focus can overlook the physician's own moral and professional responsibilities. Physicians are not mere implementers of patient wishes—they must act per their conscience and expertise. Conflicts arise when patients request futile or unethical treatments, placing physicians in moral dilemmas. However, after deeper scrutiny, patient and physician autonomy should be complementary rather than competing principles [22, 24]. Physician autonomy should be a critical factor in promoting *patient well-being* [22–24]. Autonomy enables physicians to apply their expertise in serving patient health needs in a responsible and effective manner. The ideal physician-patient relationship should be a moral partnership founded on mutual respect and a shared commitment to the patient's well-being. Both parties bring legitimate moral claims to the clinical encounter: patients contribute their values, goals, and preferences, while physicians contribute expert knowledge and moral responsibility [24], which means the physician is ethically obligated to inform and empower the patient, helping them make autonomous choices based on clear, compassionate, and clinically sound advice.

3 Impact on Physician's Autonomy

Regulatory bodies like medical chambers enforce compliance and uphold safety standards and patient-physician trust [25], with professional decision-making autonomy being crucial to these objectives. Indeed, medical ethical codes implicitly or explicitly contain physicians' professional autonomy. However, as AI systems increasingly influence clinical decision-making, the absence of clear regulatory frameworks threatens to erode professional autonomy, placing clinicians in ethically and legally ambiguous positions.

The increasing integration of AI-CDSSs into healthcare without clear regulatory guidance generates uncertainty for physicians relying on these tools [26]. The EU's AI Act introduces a risk-based framework [27] and classifies AI systems by risk level; however, it lacks specific guidance tailored to the healthcare sector [28]. Most importantly, it fails to delineate the responsibilities of healthcare providers and clinicians in ensuring compliance and managing errors, instead shifting the burden to system developers and distributors [29]. This regulatory ambiguity can undermine physician autonomy, leaving clinicians without adequate legal or institutional support when making decisions that involve AI assistance. Even the mandated human oversight for high-risk systems lacks a concrete implementation framework [26, 29].

Despite these regulatory ambiguities, discussions about AI in healthcare often emphasize its capacity to perform as well as, or even more accurately than, individual physicians in medical diagnostics [12, 30]. However, the central issue in current debates is not how to replace physicians with AI systems but how to implement these technologies into clinical practice to support and enhance medical decision-making through effective collaboration between human and artificial agents.

Central to concerns about such integration is the *black box problem* (BB), which refers to the opacity of processes by which specific ML-based systems (e.g., LLMs, deep learning) produce outputs after being trained on vast datasets [31–33]. Jenna Burrell's [32] influential typology of algorithmic opacity distinguishes *intentional opacity* when systems are deliberately kept secret (e.g., for proprietary reasons); *illiterate opacity*, which results from users' lack of technical understanding; and *intrinsic opacity*, which arises from the inherent complexity of ML models, making their decision-making processes difficult to interpret even for experts. When such systems are deployed in clinical contexts, they can undermine physicians' autonomy by limiting their ability to understand, interrogate, and challenge the rationale behind AI-generated recommendations [23, 24, 34]. In such scenarios, clinicians may feel compelled to defer to the system's outputs without fully comprehending them, effectively relinquishing meaningful control over clinical decision-making.

Beyond computational opacity (the BB problem), an additional kind of opacity emerges: the *values-black box*. While computational opacity concerns *how* AI arrives at a decision, value opacity concerns *why* it favors a particular outcome [24]. AI systems, specifically those relying on ML, can inadvertently extract and perpetuate problematic value judgments and biases within their training data [35, 36]. Furthermore, this issue intersects with what Burrell [32] identifies as "intentional opacity." Clinicians often lack insight into the values guiding the developers of AI tools, especially in healthcare, where profit motives may not align with patients' best interests or values [37]. However, the mere fact that ethical values are implicitly present in these AI systems is not inherently detrimental to the autonomy of the decision-maker. As Joshua Hatherley [38] suggests, if clinicians are aware of such ethical values and informed about the patient's preferences, the AI tool can function as a decision-support system, similar to a textbook, offering recommendations to inform a collaborative decision-making process.

It is important to note that certain aspects of the BB problem are relatively easy to address. For example, "illiterate opacity" can be mitigated through targeted education and training of clinicians, increasing their understanding of AI systems. Furthermore, "intrinsic opacity" shares similarities with a degree of accepted uncertainty in medical practice, where statistical reasoning is often employed even when underlying mechanisms remain unknown [39]. However, the values-BB presents a more profound challenge. It introduces the potential for AI systems to promote specific, potentially hidden values, which may conflict with a patient's best interest and ethical medical practice in more subtle and pervasive ways than a lack of computational transparency. A significant part of this challenge lies in the difficulty of detecting these embedded values within the system and compelling developers to be transparent about the value assumptions that have shaped their design and training processes.

Beyond opacity, physician-AI interaction introduces cognitive concerns. Clinicians may exhibit *automation bias* by over-relying on AI suggestions [40, 41].[3] Similarly, *confirmation bias* can reduce critical scrutiny when AI outputs align with initial predictions [44, 45]. Although not inherent to BB opacity, AI's lack of transparency can exacerbate these biases. Clinicians may face challenges in resolving disagreements with AI systems due to the opacity of their internal workings, which can lead to an unwarranted acceptance of the system's output, driven by fear of responsibility for errors when disagreeing with the AI [43]. Additionally, the lack of precise regulation regarding AI does not facilitate clinicians' ability to decide when and how to contest AI-generated output. Additionally, physicians may face increased pressure from healthcare institutions to accept the output of automated systems driven by the pursuit of speed, efficiency, and cost reduction. While this pressure can compromise physician autonomy, it is important to acknowledge that it is not entirely novel, as physicians frequently encounter external pressures to prescribe specific medications or adhere to particular treatment protocols.

Adding to this complex landscape is the phenomenon of clinical *deskilling*. In this phenomenon, repeated reliance on automated tools may lead to a decline in existing competencies or failure to acquire new ones [46]. It occurs when repeated reliance on automation limits active engagement in complex decisions. For instance, if an AI system consistently provides accurate recommendations, a clinician may

[3] This concern aligns with a broader issue known as *the crisis of moral patiency*, wherein individuals become passive recipients of machine decisions, relinquishing their autonomy and decision-making authority [42, 43].

gradually lose proficiency in independently interpreting diagnostic tests or formulating differential diagnoses. Over time, this can lead to a diminished ability to function without technological assistance. However, the risk of deskilling is not unique to AI or medicine; it accompanies the automation of any professional task. Its impact often depends on the user's baseline expertise and the nature of their interaction with the system.

On the positive side, AI can free up physicians' time from administrative and repetitive tasks, allowing them to dedicate more time to complex clinical decision-making and interacting with patients, which are more meaningful aspects of their profession [3, 47]. Furthermore, AI systems can serve as valuable resources for physicians, providing access to the latest research findings, aiding in literature review, and enhancing their understanding of complex cases. AI can even assist in navigating the intricate landscape of legal and ethical regulations relevant to specific medical situations.[4] However, these benefits are tempered by the need to invest time in training to use AI tools effectively, especially considering that AI systems are constantly evolving and require continuous education.

These considerations illustrate that the impact of AI on physician autonomy is multifaceted, varying in direction (positive or negative influence) and intensity (ranging from minor adjustments to significant limitations or even complete loss of autonomy). This complex interplay of factors, many of which are essentially outside the direct control of physicians, including the fundamental design of AI systems, the specific policies and protocols of the healthcare institution, as well as the physician's pre-existing level of expertise, further shape this impact. However, it is also important to recognize that physicians' proactive engagement can mitigate some of these concerns through continuous education and training on AI technologies and their ethical use.

4 Impact on Patient's Autonomy

Physician autonomy is not an end in itself but serves an instrumental role in promoting the welfare of the patient and enabling patient autonomy. When this autonomy is undermined, particularly by opaque AI systems, it may indirectly threaten the patient's own capacity for informed and autonomous decision-making. Nevertheless, the relationship between physician and patient autonomy is not strictly linear. This chapter examines how limitations on physician autonomy may influence or restrict patient autonomy and, crucially, how AI can directly impact patient autonomy, independent of the physician's role.

In this context, one of the central ethical concerns is the risk of regression to paternalism. As Xanhui Xu and Kyle M.J. Shuttleworth [49] argue, deploying BB AI systems in clinical settings can hinder transparency, making it more difficult for patients to give fully informed consent. Similarly, Berman Chan [50] notes that

[4] The development of software tools like METHAD (Medical ETHical ADvisor) exemplifies AI's potential to support physicians in addressing ethical dilemmas [48].

opacity has been increasingly associated with a revival of paternalistic dynamics in medicine, echoing concerns raised by Rosalind J. McDougall [37] and Thomas Grote and Philipp Berens [51]. When clinicians defer to AI outputs without critical evaluation or fail to communicate the reasoning behind a recommendation, they may inadvertently position the AI as a clinical authority. This dynamic can marginalize the patient's role in decision-making and shift the patient-physician relationship from a dialogue to one of compliance.

Consequently, *transitive paternalism* arises when physicians accept AI outputs without discussion and further scrutiny, presenting them to the patient as preordained conclusions rather than as part of a deliberative process. The perception of AI outputs as statistically superior or authoritative can reduce a clinician's willingness, or perceived ability, to invite patient participation. As a result, patients may feel pressured to accept recommendations that were not negotiated meaningfully with them.

Thus, opacity not only affects clinicians' interpretive capacities; it also creates challenges for patient understanding. A key issue here is the physician's inability to explain the basis of the AI recommendation. Although the inability to explain complex models may undermine informed consent, this can be overstated. If physicians clarify what is and is not knowable about the system's reasoning and rationale for its use, they can preserve a certain level of transparency. The real problem, then, is not just the *technical opacity* of AI but the *relational opacity* that arises when patients are left uninformed about how decision-making procedures work.

From the patient's perspective, Burrell's [32] typology does not capture all forms of opacity. First, patients' general lack of medical literacy creates a *double illiterate opacity*, as they are unfamiliar with both AI's technical dimensions and the medical knowledge underpinning clinical judgment. Second, a subtler opacity exists in the physician's decision-making process itself: patients rarely receive information about how conclusions were reached, including uncertainties about the use of AI tools and the reasons behind them. This communicative gap creates fertile ground for real paternalism, as patients lack both raw data and the context they need to understand the conclusion. Ultimately, patient autonomy must be protected not primarily from the technical complexity of AI but from this breakdown in clinician-patient transparency. Therefore, it is not the inscrutability of the algorithm itself that threatens the patient's autonomy but rather the failure to integrate the patient's perspective into the clinical reasoning process.

The traditional problem of informed consent centers on how much information should be disclosed to the patient. Regarding the disclosure of risk, it is generally accepted that it should not exceed the boundaries of *reasonableness*. Nevertheless, there is no universally accepted interpretation of this standard. Without delving into more profound philosophical questions on this concept, we can refer to the two prevailing interpretations encountered in clinical practice: the patient-centered standard (what a reasonable person in the patient's position would consider a risk worth disclosing) and the physician-centered standard (what a reasonable physician would disclose under similar circumstances) [52]. In the context of AI-CDSSs, the central issue shifts to whether patients should be informed that such a system played a role in the diagnostic process, particularly if its use introduces a material risk, such as errors due to algorithmic opacity or bias, that a reasonable patient might want to be aware of. Given the BB problem, this appears to be a material piece of information, as some scholars argue [53], as it raises concerns about opacity and accountability.

An empirical study by Hai J. Park [54] found that patients valued knowing AI was used in diagnosis regardless of risk, widespread adoption, or superior performance, challenging assumptions about disclosure in low-risk contexts. The study highlights that such information, significant for informed consent, supports broadening the patient-centered standard to include new information about the diagnostic process. The WHO [55] echoes this, recommending the inclusion of AI use in consent procedures and warning that failure to do so risks undermining autonomy and public trust. Legal frameworks, such as the GDPR [56], reinforce this by granting individuals the right to meaningful information about impactful automated decisions, although excluding those involving significant human input [57]. However, the distinction between decision support and full automation [57] creates a grey area where human oversight may lessen transparency duties and demands further scrutiny.

Simply informing patients about AI use (the *disclosure thesis*) is insufficient to safeguard autonomy. Critique notes [38] that overemphasis on disclosure risks reducing it to a symbolic gesture, as it allows clinicians, hospitals, or AI developers to deflect accountability when harm results from the inappropriate use of such systems.[5] In these cases, the core issue is not merely a failure to inform but an overreliance on systems whose outputs may be poorly understood or misapplied. Safeguarding autonomy, therefore, requires more than disclosure; it demands a shared responsibility among all stakeholders [38, 51]. AI developers must design systems with explainability and contextual appropriateness in mind. Healthcare institutions should implement protocols that ensure AI supports, rather than supplants, clinical judgment. Clinicians, in turn, must critically engage with AI outputs and incorporate them into transparent, patient-centered decision-making. Only through such coordinated efforts can the autonomy of both physicians and patients be meaningfully upheld in the age of algorithmic medicine.

While some argue that bioethical discourse often reduces patient autonomy to procedural informed consent [59], emerging technologies like wearables, mental health chatbots, and AI surrogates mandate a broader discussion of autonomy, one that entails a level of independence from the informational input provided by healthcare professionals in a clinical setting. AI can enhance patient autonomy by providing personalized health information and recommendations while also increasing access to healthcare.

In wearable health technologies, AI's transformative influence on patient access to personal health information is evident. By integrating ML algorithms and smart materials, health wearables now collect a diverse range of physiological parameters

[5] This concern is closely related to the well-documented problem of *responsibility gaps* in AI, where the diffuse nature of AI development and deployment can make it challenging to attribute clear moral or legal responsibility when AI systems cause harm [58].

and biomarkers (e.g., heart rate, glucose), enabling personalized health analytics and predictive insights [60]. Beyond wearables, AI also plays a vital role in addressing mental health needs and increasing access to mental health support. The growing demand for psychotherapy and a scarcity of available professionals make AI particularly relevant in expanding mental health support. AI mental health chatbots have significant potential to increase access to psychotherapeutic care and psychoeducation, especially for individuals with limited access to traditional services. Their constant availability and the option for anonymous interaction can empower patients to proactively seek help and engage more actively in managing their mental health. Research indicates, however, that while chatbots are successful in alleviating conditions such as anxiety, they are not necessarily more effective than in-person therapists [61]. Together, these examples point to a shift in how patients' autonomy is exercised—from being merely reactive to medical authority toward active engagement with health through personalized, AI-mediated tools.

Furthermore, AI holds significant potential in surrogate decision-making and advanced care planning. Surrogate decision-makers are typically close family members who best reflect the patient's preferences, understand their values, and often serve as primary caregivers [62]. However, emerging AI systems trained on a patient's digital footprint, prior health choices, and personal data present a compelling case for serving as surrogate decision-makers in specific contexts [63, 64]. If designed with sufficient ethical safeguards, AI surrogates could surpass human proxies in consistency, memory of patient preferences, and resistance to emotional bias. For instance, the idea is to utilize AI tools to support advanced care planning by inferring patient preferences from structured data and clinical histories [65]. Significantly, such tools could expand patient autonomy by extending a form of self-directedness even when individuals can no longer make decisions themselves [66]. By encoding a patient's long-term values and treatment inclinations, AI surrogates may enable decisions that are not only informed but also deeply personal, thereby reducing reliance on potentially fallible or conflicted human proxies. Nonetheless, concerns regarding data privacy, value misalignment, and system opacity require continued scrutiny and careful regulation.

5 Conclusion

The rise of AI in healthcare has a complex impact on the autonomy of physicians and patients, as well as the relationship between them. From the physician's perspective, it can enhance autonomy through decision support or undermine it via opaque algorithms and unclear regulations. AI-CDSSs may limit physicians' ability to interpret, question, or override recommendations, particularly in the face of the BB problem, which can gradually shift their role from decision-makers to system operators. Regulatory uncertainty and institutional incentives to adopt AI quickly (often without sufficient training or safeguards) further challenge clinicians' ability to exercise independent, context-sensitive judgment.

This erosion of physician autonomy often undermines patient autonomy by weakening the process of shared decision-making. When physicians defer to AI without scrutiny, they risk presenting recommendations as final decisions, limiting patient participation, and reverting to a paternalistic model. However, patient autonomy can also be compromised independently, for example, when patients are unaware of decision-making procedures. Autonomy loss here stems not only from technical opacity but also from failures in communication and engagement.

Despite these challenges, AI also holds the potential to empower both clinicians and patients. For physicians, AI can streamline administrative tasks, enhance diagnostic accuracy, and allow more time for patient interaction. For patients, wearables, chatbots, and AI-based health insights provide new avenues for engagement and self-care. These innovations allow autonomy to move beyond traditional informed consent toward active, ongoing health involvement.

Still, autonomy gains are not guaranteed. The presence of hidden value judgments in AI systems (the values-BB) means that even seemingly neutral tools may subtly influence decisions. Simply disclosing AI involvement is not enough; clinicians and developers must take responsibility for how these tools are used and understood. In this context, the principle of explicability, combining transparency in system design with accountability for outcomes, emerges as a necessary complement to traditional bioethical principles, especially autonomy. Upholding explicability ensures that AI contributes to, rather than detracts from, meaningful and ethically sound clinical decision-making.

Although physician autonomy supports patient autonomy, the two can diverge. Patient autonomy may suffer even when physician autonomy is intact, or vice versa. For instance, a clinician might retain full decision-making authority but fail to communicate clearly with the patient about the role of AI, thereby excluding them from meaningful participation in the decision-making process. Conversely, patients may gain autonomy even when physicians feel constrained by opaque or institutionally mandated AI systems. Moreover, AI can directly impact patients without physician mediation through the use of mental health chatbots, wearables, or AI surrogates. While these tools *may* enhance patient autonomy by increasing access, personalization, and continuity of care, such benefits are contingent upon their alignment with the principles of transparency, value awareness, and ethical design, emphasized above.

In conclusion, patient well-being and autonomy are among the highest values in healthcare and must not be sidelined. Patient interests must always be a priority in cases of divergence between physician and patient autonomy. AI tools in clinical decision-making must serve the patient.

Acknowledgment The authors utilized Google Gemini Advanced and ChatGPT (OpenAI, GPT-4) to support the editorial refinement of the manuscript, including sentence restructuring, paraphrasing, and summarising specific sections. AI tools were not used to conduct original research, generate scientific content, formulate arguments, or create citations. All intellectual content remains the sole responsibility of the authors. Outputs modified with the assistance of AI have been critically reviewed and edited to ensure compliance with academic standards and research integrity.

Competing Interests This chapter is the result of research conducted at the University of Belgrade—Faculty of Philosophy, which was financially supported by the Ministry of Science, Technological Development, and Innovations of the Republic of Serbia (451-03-137/2025-03/200163).

References

1. Sutton RT, Pincock D, Baumgart DC, Sadowski DC, Fedorak RN, Kroeker KI (2020) An overview of clinical decision support systems: benefits, risks, and strategies for success. NPJ Digit Med 3:17. https://doi.org/10.1038/s41746-020-0221-y
2. Bitkina OV, Park J, Kim HK (2023) Application of artificial intelligence in medical technologies: a systematic review of main trends. Digit Health 9. https://doi.org/10.1177/20552076231189331
3. Lombi L, Rossero E (2024) How artificial intelligence is reshaping the autonomy and boundary work of radiologists. A qualitative study. Soc Sci Med 46(2):200–218. https://doi.org/10.1111/1467-9566.13702
4. Singhal K, Azizi S, Tu T, Mahdavi SS, Wei J, Chung HW et al (2023) Large language models encode clinical knowledge. Nature 620(7973):172–180. https://doi.org/10.1038/s41586-023-06291-2
5. Mendel T, Singh N, Mann DM, Wiesenfeld B, Nov O (2025) Laypeople's use of and attitudes toward large language models and search engines for health queries: survey study. J Med Internet Res 27:e64290. https://doi.org/10.2196/64290
6. Wang L, Wan Z, Ni C, Song Q, Li Y, Clayton E, Malin B, Yin Z (2024) Applications and concerns of ChatGPT and other conversational large language models in health care: systematic review. J Med Internet Res 26:e22769. https://doi.org/10.2196/22769
7. OpenAI (2024) GPT-4 technical report (Version 6). Preprint at https://arxiv.org/abs/2303.08774
8. Kassirer JP, Wong JB, Kopelman RI (2010) Learning clinical reasoning, 2nd edn. Lippincott Williams & Wilkins, Philadelphia
9. National Academies of Sciences, Engineering, and Medicine (2015) Improving diagnosis in health care. The National Academies Press, Washington, DC. https://doi.org/10.17226/21794
10. Sittig DF, Wright A, Middleton B (2016) Clinical decision support: a 25-year retrospective and a look ahead. Yearb Med Inform 25(S 01):S103–S116. https://doi.org/10.15265/IYS-2016-s034
11. Yu KH, Beam AL, Kohane IS (2018) Artificial intelligence in healthcare. Nat Biomed Eng 2(10):719–731. https://doi.org/10.1038/s41551-018-0305-z
12. Topol EJ (2019) High-performance medicine: the convergence of human and artificial intelligence. Nat Med 25:44–56. https://doi.org/10.1038/s41591-018-0300-7
13. Chen J, Miao C (2025) DeepSeek deployed in 90 Chinese tertiary hospitals: how artificial intelligence is transforming clinical practice. J Med Syst 49:53. https://doi.org/10.1007/s10916-025-02181-4
14. Beauchamp TL, Childress JF (2019) Principles of biomedical ethics, 8th edn. Oxford University Press, New York
15. Evans D (2009) Article 5: autonomy and individual responsibility. In: ten Have HAMJ, Jean MS (eds) The UNESCO universal declaration on bioethics and human rights: background, principles and application. UNESCO Publishing, Paris, pp 61–66. Available at https://unesdoc.unesco.org/ark:/48223/pf0000179844. Accessed 1 May 2025
16. Floridi L, Cowls J (2021) A unified framework of five principles for AI in society. In: Floridi L (ed) Ethics, governance, and policies in artificial intelligence. Springer, Cham, pp 5–17. https://doi.org/10.1007/978-3-030-81907-1_2
17. Raso F, Hilligoss H, Krishnamurthy V, Bavitz C, Kim LY (2018) Artificial intelligence & human rights: opportunities & risks. Berkman Klein Center Research Publication No. 2018-6. Available at https://ssrn.com/abstract=3259344. Accessed 15 May 2025

18. Young R (2009) Informed consent and patient autonomy. In: Kuhse H, Singer P (eds) A companion to bioethics, 2nd edn. Wiley-Blackwell, Chichester, pp 530–540. https://doi.org/10.1002/9781444307818.ch44
19. Shah P, Thornton I, Kopitnik NL et al (2024) Informed consent. In: StatPearls. StatPearls Publishing, Treasure Island. Available at https://www.ncbi.nlm.nih.gov/books/NBK430827/. Accessed 25 May 2025
20. Council of Europe (1997) Convention for the protection of human rights and dignity of the human being with regard to the application of biology and medicine: convention on human rights and biomedicine (Oviedo Convention). Available at https://rm.coe.int/168007cf98. Accessed 25 May 2025
21. Capron AM (2009) Advance directives. In: Kuhse H, Singer P (eds) A companion to bioethics, 2nd edn. Wiley, Chichester, pp 299–311. https://doi.org/10.1002/9781444307818.ch26
22. Pellegrino ED (1994) Patient and physician autonomy: conflicting rights and obligations in the physician–patient relationship. J Contemp Health Law Policy 10(1):47–68. https://scholarship.law.edu/jchlp/vol10/iss1/8
23. Funer F, Wiesing U (2024) Physician's autonomy in the face of AI support: walking the ethical tightrope. Front Med 11:1324963. https://doi.org/10.3389/fmed.2024.1324963
24. Lorenzini G, Arbelaez Ossa L, Shaw DM, Elger BS (2023) Artificial intelligence and the doctor–patient relationship: expanding the paradigm of shared decision making. Bioethics 37(4):424–429. https://doi.org/10.1111/bioe.13158
25. World Medical Association (2009) WMA Declaration of Madrid on Professionally-Led Regulation. Adopted by the 60th WMA General Assembly, New Delhi, India, October 2009 and revised by the 70th WMA General Assembly, Tbilisi, Georgia, October 2019. Available at https://www.wma.net/policies-post/wma-declaration-of-madrid-on-professionally-led-regulation. Accessed 24 Mar 2025
26. Smith H, Downer J, Ives J (2024b) Clinicians and AI use: where is the professional guidance? J Med Ethics 50(7):437–441. https://doi.org/10.1136/jme-2022-108831
27. Regulation (EU) 2024/1689 of the European Parliament and of the Council of 13 June 2024 laying down harmonised rules on artificial intelligence and amending Regulations (EC) No 300/2008, (EU) No 167/2013, (EU) No 168/2013, (EU) 2018/858, (EU) 2018/1139 and (EU) 2019/2144 and Directives 2014/90/EU, (EU) 2016/797 and (EU) 2020/1828 (Artificial Intelligence Act). Off J Eur Union L 2024/1689, 12 July 2024
28. van Kolfschooten H, van Oirschot J (2024) The EU Artificial Intelligence Act (2024): implications for healthcare. Health Policy 149:105152. https://doi.org/10.1016/j.healthpol.2024.105152
29. Vardas EP, Marketou M, Vardas PE (2025) Medicine, healthcare and the AI Act: gaps, challenges and future implications. Eur Heart J Digit Health:ztaf041. https://doi.org/10.1093/ehjdh/ztaf041
30. Topol EJ (2024) Toward the eradication of medical diagnostic errors. Science 383:eadn9602. https://doi.org/10.1126/science.adn9602
31. Berber A, Srećković S (2023) When something goes wrong: who is responsible for errors in ML decision-making? AI Soc. https://doi.org/10.1007/s00146-023-01640-1
32. Burrell J (2016) How the machine 'thinks': understanding opacity in machine learning algorithms. Big Data Soc 3(1):1–12. https://doi.org/10.1177/2053951715622512
33. Wadden J (2022) Defining the undefinable: the black box problem in healthcare artificial intelligence. J Med Ethics 48:764–768
34. Smith H, Birchley G, Ives J (2024a) Artificial intelligence in clinical decision-making: rethinking personal moral responsibility. Bioethics 38:78–86. https://doi.org/10.1111/bioe.13222
35. O'Neil C (2016) Weapons of math destruction: how big data increases inequality and threatens democracy. Crown, New York
36. Mehrabi N, Morstatter F, Saxena N, Lerman K, Galstyan A (2022) A survey on bias and fairness in machine learning. ACM Comput Surv (CSUR) 54(6):115. https://doi.org/10.1145/3457607

37. McDougall RJ (2019) Computer knows best? The need for value-flexibility in medical AI. J Med Ethics 45(3):156–160. https://doi.org/10.1136/medethics-2018-105118
38. Hatherley J (2024) Are clinicians ethically obligated to disclose their use of medical machine learning systems to patients? J Med Ethics. https://doi.org/10.1136/jme-2024-109905
39. London AJ (2019) Artificial intelligence and black-box medical decisions: accuracy versus explainability. Hastings Cent Rep 49(1):15–21. https://doi.org/10.1002/hast.973
40. Goddard K, Roudsari A, Wyatt JC (2012) Automation bias: a systematic review of frequency, effect mediators, and mitigators. J Am Med Inform Assoc 19(1):121–127. https://doi.org/10.1136/amiajnl-2011-000089
41. Kordzadeh N, Ghasemaghaei M (2022) Algorithmic bias: review, synthesis, and future research directions. Eur J Inf Syst 31(3):388–409. https://doi.org/10.1080/0960085X.2021.1927212
42. Danaher J (2019) The rise of the robots and the crisis of moral patiency. AI Soc 34:129–136. https://doi.org/10.1007/s00146-017-0773-9
43. Berber A (2023) Automated decision-making and the problem of evil. AI Soc. https://doi.org/10.1007/s00146-023-01814-x
44. Mendel R, Traut-Mattausch E, Jonas E, Leucht S, Kane JM, Maino K, Kissling W, Hamann J (2011) Confirmation bias: why psychiatrists stick to wrong preliminary diagnoses. Psychol Med 41(12):2651–2659. https://doi.org/10.1017/S0033291711000808
45. Elston DM (2020) Confirmation bias in medical decision-making. J Am Acad Dermatol 82(3):572
46. Natali C, Marconi L, Dias Duran LD, Miglioretti M, Cabitza F (2025) AI-induced deskilling in medicine: a mixed method literature review for setting a new research agenda. Preprint at https://ssrn.com/abstract=5166364
47. Gandhi TK, Classen D, Sinsky CA, Rhew DC, Vande Garde N, Roberts A, Federico F (2023) How can artificial intelligence decrease cognitive and work burden for front line practitioners? JAMIA Open 6(3):ooad079. https://doi.org/10.1093/jamiaopen/ooad079
48. Meier LJ, Hein A, Diepold K, Buyx A (2022) Algorithms for ethical decision-making in the clinic: a proof of concept. Am J Bioeth 22(7):4–20. https://doi.org/10.1080/15265161.2022.2040647
49. Xu H, Shuttleworth KMJ (2024) Medical artificial intelligence and the black box problem: A view based on the ethical principle of "do no harm". Intell Med 4(1):52–57. https://doi.org/10.1016/j.imed.2023.08.001
50. Chan B (2023) Black-box assisted medical decisions: AI power vs. ethical physician care. Med Health Care Philos 26:285–292. https://doi.org/10.1007/s11019-023-10153-z
51. Grote T, Berens P (2020) On the ethics of algorithmic decision-making in healthcare. J Med Ethics 46(3):205–211. https://doi.org/10.1136/medethics-2019-105586
52. Cohen IG (2020) Informed consent and medical artificial intelligence: what to tell the patient? Geo LJ 108:1425–1469
53. Kiener M (2021) Artificial intelligence in medicine and the disclosure of risks. AI Soc 36(2):705–713. https://doi.org/10.1007/s00146-020-01085-w
54. Park HJ (2024) Patient perspectives on informed consent for medical AI: a web-based experiment. Digit Health 10:20552076241247938. https://doi.org/10.1177/20552076241247938
55. World Health Organization (2021) Ethics and governance of artificial intelligence for health: WHO guidance. World Health Organization, Geneva. Available at https://iris.who.int/handle/10665/341996. Accessed 8 Apr 2025
56. Regulation (EU) 2016/679 of the European Parliament and of the Council of 27 April 2016 on the protection of natural persons with regard to the processing of personal data and on the free movement of such data, and repealing Directive 95/46/EC (General Data Protection Regulation). Off J Eur Union L 119:1–88. https://eur-lex.europa.eu/eli/reg/2016/679/oj
57. Pruski M (2024) AI-enhanced healthcare: not a new paradigm for informed consent. J Bioeth Inq 21:475–489. https://doi.org/10.1007/s11673-023-10320-0
58. Berber A, Mijić J (2024) Understanding moral responsibility in automated decision-making: responsibility gaps and strategies to address them. Theoria 67(3):177–192. https://doi.org/10.2298/THEO2403177B

59. O'Neill O (2002) Autonomy and trust in bioethics. Cambridge University Press, Cambridge
60. Chen S, Fan S, Qiao Z, Wu Z, Lin B, Li Z, Riegler MA, Wong MYH, Opheim A, Korostynska O, Nielsen KM, Glott T, Martinsen ACT, Telle-Hansen VH, Lim CT (2025) Transforming healthcare: intelligent wearable sensors empowered by smart materials and artificial intelligence. Adv Mater 37(25):2500412. https://doi.org/10.1002/adma.202500412
61. Spytska L (2025) The use of artificial intelligence in psychotherapy: development of intelligent therapeutic systems. BMC Psychol 13:175. https://doi.org/10.1186/s40359-025-02491-9
62. Brock DW (2009) Medical decisions at the end of life. In: Kuhse H, Singer P (eds) A companion to bioethics, 2nd edn. Wiley, Chichester, pp 263–273. https://doi.org/10.1002/9781444307818.ch23
63. Rid A, Wendler D (2014) Use of a patient preference predictor to help make medical decisions for incapacitated patients. J Med Philos 39(2):104–129
64. Earp BD, Porsdam Mann S, Allen J, Salloch S, Suren V, Jongsma K, Braun M, Wilkinson D, Sinnott-Armstrong W, Rid A, Wendler D, Savulescu J (2024) A personalized patient preference predictor for substituted judgments in healthcare: technically feasible and ethically desirable. Am J Bioeth 24(7):13–26. https://doi.org/10.1080/15265161.2023.2296402
65. Patel M, Mara A, Acker Y, Gollon J, Wolf S, Zafar Y, LeBlanc TW, Ma J (2024) Using machine learning for targeted advance care planning (ACP) conversations in patients with cancer: a quality improvement initiative. J Clin Oncol 42(16_suppl):1552–1552. https://doi.org/10.1200/JCO.2024.42.16_suppl.1552
66. Earp BD, Porsdam Mann S, van Veenendaal T, Allen JW, Salloch S, Jongsma KR, Braun M, Sinnott-Armstrong W, Savulescu J, Rid A, Wendler D (2025) Respecting formerly autonomous persons: clarifying the role of the Personalised Patient Preference Predictor (P4) in substituted judgement. J Med Ethics. https://doi.org/10.1136/jme-2025-110729

Andrea Berber is a Research Associate at the University of Belgrade. She obtained her PhD in philosophy from the University of Belgrade in 2019. Andrea has published on the philosophy of artificial intelligence in journals such as *Minds & Machines* and *AI & Society*. Her current research focuses on the epistemic and ethical challenges raised by the use of AI-based tools in decision-making and creative endeavors. Andrea's research interests include the philosophy and ethics of AI, social epistemology, and the implications of emerging technologies on human autonomy and responsibility.

Jelena Mijić is an Assistant Professor at the Faculty of Philosophy, University of Belgrade, in the Department of Philosophy. She holds a PhD in philosophy from the University of Belgrade. Jelena's main areas of interest are epistemology and action theory. Drawing on these philosophical disciplines, she investigates problems in applied ethics and epistemology, with a particular focus on issues related to artificial intelligence. Her work aims to bridge theoretical philosophy and practical challenges posed by AI technologies, especially concerning moral agency and epistemic practices.

AI in Clinical Practice: A Feyerabendian Pluralistic Model for Physician-Patient Relationships

11

Deivide Garcia da Silva Oliveira

Abstract

As AI technology extends through all areas, some say that we are entering an AI-driven world, one with a deterministic worldview and with no need for understanding why and how we choose/believe the way we do. Accordingly, it is a fact that AI disruption has advanced throughout our individual lives, the economy, and various professions, including medicine. As this deterministic world expands, it integrates AI into medical decision-making, affecting the autonomy of both physicians and patients. Epistemic and ethical implications can also be identified, particularly, concerning of AI's influence on the physician-patient relationship. As AI recommendations pressure patients and physicians, tension emerges between AI-driven recommendations and the physician's convincing powers. This chapter aims to explore the disruptive nature of AI in terms of the dynamics and trustworthiness typical of a physician-patient relationship. To this aim, Emanuel and Emanuel's traditional four mainstream models of that relationship have been scrutinized, and the question of which model would best suit our times explored. Given the complexity of this relationship, coupled with AI, a new model, the *Philosophically Pluralistic Model*, inspired in Feyerabend's philosophy, is finally outlined to better address the ethical, scientific, and epistemic issues surrounding the disruption of AI in the physician-patient relationship.

Keywords

Pluralistic model · Emergency-paced scenarios · AI-physician-patient relationship

D. G. da Silva Oliveira (✉)
Universidade Federal de Sergipe, Sergipe, Brazil

M. Anzalone et al. (eds.), *Reconfiguring Human Autonomy*, Integrated Science 40,
https://doi.org/10.1007/978-3-032-14537-6_11

1 Introduction: AI's Impact on Autonomy and Ethics

As artificial intelligence increasingly becomes integrated into our lives, the medical field is not immune to its effects. Clinical environments are prime examples of how AI can disrupt established norms regarding medical authority, patient autonomy, and ethical decision-making. While medical settings have always been subject to scrutiny, the introduction of AI heightens the sensitivity of these already complex situations and the decisions surrounding them. Concerning its impact on humans, we should also make a preliminary distinction regarding the type of AI being discussed.

According to Mustafa Suleyman (DeepMind co-founder), we live in an ACI age (artificial capable intelligence, which is similar to weak AI), where AI achieves "a wide range of complex tasks but is still a long way from being fully general" [1 , p. vii]. In this sense, we have weak AIs that can analyze images, like healthy or carcinogenic ones, with great precision (ACI). Suleyman also points out that we are still a long way from replacing physicians with AIs (such as strong AI or just AGI: artificial general intelligence) that greatly exceed the smartest human. That being said, more dangerous problems, such as physicians being replaced by AI, are not here yet. However, despite the fact that we are living in an ACI age, many problems have already emerged. For instance, whose medical advice do patients trust more, that given by AI or by physicians? According to Ryosuke Yokoi and colleagues, today's patients indeed trust physicians more than AI [2], but this could change, and it may occur at varying paces across different areas of medicine. What are companies likely to do if we reach an AI that could act like a good general physician? Furthermore, what is likely to happen for those who cannot afford human doctors, whereas seeing AI doctors drastically decreases their costs while offering robust diagnoses? In a world where predictive algorithms increasingly guide diagnosis and treatments, what happens to the physician-patient relationship when both are influenced—sometimes invisibly—by AI? To dive into these questions, I will review four existing models of the physician-patient relationship, as proposed by Ezekiel J. Emanuel and Linda L. Emanuel [3]. I will also argue that, although Emanuel and Emanuel's work offers valuable perspectives, it lacks the conceptual tools required to address AI's disruptive impact for physician-patient relationships.

This chapter aims to provide a model that can be applied in medical practice within the emerging AI-driven landscape. To this aim, a Philosophically Pluralistic Model (PhiPlur) is outlined. I argue that the physician-patient relationship in our times requires a new framework that recognizes not only the technical benefits of AI but also its epistemic and ethical limitations. To develop such a model, I have drawn inspiration from Feyerabend's pluralism, while incorporating recent critiques of AI capabilities. The PhiPlur framework addresses both high-pressure medical situations and reflective care contexts, considering the implications of an AI-driven world. As a result, a shift in the traditional goals of the paternalistic model (Pat-M) is proposed, advocating for a focus on minimizing or eliminating patients' suffering rather than solely prioritizing healing, although when healing is realistic, it also serves the reduction of suffering. This change recognizes that healing is not always a viable or a priority option for all patients. Consequently, the PhiPlur model effectively preserves human agency while addressing the complex relationships among data, values, and patient suffering in the age of AI.

The chapter is structured as follows: In Sect. 2, Emanuel and Emanuel's [3] four models of physician-patient relationships are briefly reviewed. In Sect. 3, the AI's impact on each of the four models is explored. In Sect. 4, I argue that one of those models, although unsatisfactory, is better suited for our new world, and then, drawing from it and the outlined discussion, I introduce a new model, inspired by Feyerabend's pluralism, and some key features identified in Emanuel and Emanuel's four models. Section 5 presents some concluding remarks.

2 Revisiting and Evaluating the Emanuel and Emanuel Models

With their paper Emanuel and Emanuel [3] enter a well-developed debate—flourishing at least since 1972. Four models of the physician-patient interaction are thereby outlined according to different understandings of autonomy and values in that interaction. Furthermore, one should add that—regardless of whether these models are partially or entirely replaced by new ones—the driving question behind these models is still thriving and informing most models, new or old, even today. With AI entering the game, the same question has increased in importance and acquired a new angle. Emanuel and Emanuel ask: What should be the ideal physician-patient relationship? For the purposes of this chapter, this can be reformulated as: What is the ideal physician-AI-patient interaction? Emanuel and Emanuel argue that the following four models suitably frame an answer for that question: *1-The Paternalistic Model, 2-The Informative Model, 3-The Interpretative Model, and 4-The Deliberative Model.* Each model conveys one perspective over the physician obligation, the patient values, the physician limits, and the patient autonomy [3, p. 2221]. In this respect, it is important to emphasize that a complicating factor currently interferes with any applied model. The presence of AI in medicine and in the physician's office has, potentially, both helpful and harmful effects. AI has the potential to function in different ways, such as a bridge between the physician and the patient, or between some algorithmic personal tailored medical knowledge and the physician, or between some algorithmic personal tailored medical knowledge and the patient. However, before discussing how and why each of these four models can benefit or suffer from the integration of AI, let us take a closer look at each model.

2.1 The Paternalistic Model (Pat-M)

In the authors' account, the paternalistic model—also called the parental or priestly model—aims to ensure that "patients receive the interventions that best promote their health and well-being." [3]. Prima facie, the model does not seem to be doing anything wrong. To the aim of promoting patient health and well-being, physicians use their skills to "determine" the patient's condition, disease, stage of the disease, medical tests, and treatments. How so? Based on Pat-M, "the physician presents the patient with selected *information* that will encourage the patient to consent to the intervention the physician considers best". What if the patient is still not convinced or refuses to initiate the intervention? Well, "at the extreme, the physician authoritatively informs the patient when the intervention will be initiated". Two main assumptions can be identified in this model. First, the cure of the patient is the priority. Second, it is assumed by the physician that their best medical knowledge about a disease is the same as what is best for the patient, not only as a patient but as a person in their entire complexity. It is only the physician who "can discern what is in the patient's best interest" [3]. Thus, the physician must act, and the patient will thank them; if not now, surely the patient will do it later.

2.2 The Informative Model (InF-M)

The authors inform us that the informative model is also referred to as the scientific, engineered, or consumer model. In this model, the aim for the physician is "to provide the patient with all relevant information…and to execute the selected interventions", while for the patient is "to select the medical interventions" wanted [3]. To this end, the physician gives the patient relevant information about the disease, diagnostics, therapeutics, risks, benefits, "and any uncertainties of knowledge". In the case of the InF-M, it is "the physician's obligation to provide all the available facts, and the patient's values then determine what treatments are to be given". By all available facts, the authors mean "all relevant information", which, in extreme cases, leads physicians to give the patients "all medical information relevant to their disease and available interventions" [3]. In other words, not only some selected information but all information.

2.3 The Interpretive Model (InT-M)

The authors state that the Interpretive Model aims "to elucidate the patient's values and what he or she actually wants, and to help the patient select the available medical interventions that realize these values." [3]. In other words, this model assigns physicians the task of interpreting what patients really value. The term "interpret" comes from the Latin *interpres*, meaning the person who negotiates between two parts. And why is this role needed? The model assumes that the "patient's values are not necessarily fixed and known to the patient" [3, p. 2222]. That is why the physician is required to be the translator of the values held by the patient, even though sometimes they do not really know. The physician will also be the translator of the patient's interpretations and aims related to those values. Based on the InT-M, the physician acts as a counselor, supplier of information, translating values, and suggesting fitting medical treatments. To this model, the concept of "patient autonomy is self-understanding" [3], a process undertaken with the physician's guidance.

2.4 The Deliberative Model (DeL-M)

The authors say that the Deliberative Model (DeL-M) aims "to help the patient determine and to choose the best health-related values that can be realized in the clinical situation." [3]. In this sense, considering the possible and available choices at hand, the physician acts as a friend through a dialogue on what choice would be best. To achieve this end, the physician presents what they consider the needed information to be "on the patient's clinical situation and then help elucidate the types of values embodied in the available options". Naturally, an unfolding question is this: based on what criteria such help and elucidation would rest? One answer is that these criteria are based on the physician's objectives, which are twofold. First, it is the physician's goal to suggest "why certain health-related values are more worthy" and, based on their assumption of what is worthy and what is not, the physician takes a second step and suggests why certain values "should be aspired to". However, what if the patient disagrees with the physician about what is worthy and what values should be pursued among those that are worthy? In these cases, at "the extreme, the physician and patient engage in deliberation", which, of course, for the physician, must be limited to discussing "only health-related values, that is, values that affect or are affected by the patient's disease and treatments" [3].

3 Determining AI's Potential Impact on Each Model

Emanuel and Emanuel emphasize a longstanding tension in medicine: a "struggle over the patient's role in medical decision making," often framed as a conflict "between autonomy and health, between the values of the patient and the values of the physician" [3, p. 2221]. While they trace this back to the late twentieth century, the struggle is much older. In eighteenth and nineteenth-century Vienna, at the Allgemeine Krankenhaus, puerperal (childbed) fever devastated parturient women. The hospital offered two divisions: the First, staffed by physicians and medical students, and the Second, by midwives. Despite equal delivery volumes, "an average of 600–800 mothers died each year" in the First Division, while only "about 60 deaths" occurred in the Second [4, p. 97]. Yet, when poor women attempted to avoid the First Division, they were systematically ignored. Their preferences were overruled by physician authority. This episode exemplifies how institutional power has historically overridden patient autonomy, especially when it challenges professional or epistemic authority. Today, AI introduces a new layer to this power dynamic. As patients and physicians both encounter algorithmic recommendations, patients and physicians alike should ask who/what actually holds the power over the decisions made. But to better address that question, let us revisit the four models to see how the integration of AI could alter or challenge each model's assumptions and dynamics.

3.1 The Paternalistic Model (Pat-M) Under AI

As previously stated, Pat-M's aim is to promote the patient's health and well-being, and to that end, physicians use their skills to "determine" the patient's condition, disease, stage of the disease, medical tests, and treatments [3, p. 2221]. Considering that AI now enters the equation, there are constructive and destructive ways for that determination to happen. In the Pat-M, the presence of AI could enhance the physician's diagnostic and prognostic analysis, i.e., reinforcing the physician's confidence in selecting the "best" intervention. In other words, here, AI would only make the Pat-M stiffer, the physician's powers amplified, and patients more vulnerable. To be clear, depending on the confidence of physicians in AI—backed by enough cases of allegedly successful help—the physician's critical reflection could be reduced as they perceive AI outputs as infallible—or less fallible than themselves—because of "the guise of technological neutrality and scientific objectivity, which is an added reason why we trust them." [5, p. 283]. To sum up, if the physician chose the blend of Pat-M and AI, the decisions made would not be centered on the patient, resulting in the entrenchment of a techno-clinical authority that justifies overriding patient preferences under the guise of algorithmic objectivity. That danger is even greater when the physician is not well prepared. For instance, recently, in Brazil it was reported that over 50,000 physicians graduated between 2013 and 2023 from universities whose medical programs received grades below the minimum required for approval, highlighting the associated risks [6].

3.2 The Informative Model (InF-M) Under AI

In this model, the aim of the physician is "to provide the patient with all relevant information" while the patient "selects the medical interventions" wanted [3, p. 2221]. Clearly, AI would be helpful in the informative task—for instance, providing rapid access to extensive up-to-date medical data—especially in public hospital settings where the number of patients is high. Nonetheless, it is also very easy to foresee the challenge of information overload and difficulty in curating "relevant" information for each patient. As a result, patients would get less clarity than just a caring physician could provide, compromising patient autonomy far beyond what just a physician-patient relationship could do. Thus, an "invisible layer of technological mediation, which structures and influences our decision-making, renders us radically susceptible to manipulation. At an individual level, it threatens our autonomy, […], which [is] premised on individuals being capable of independent decision-making." [7, p. 404]. Patients may become increasingly susceptible to misunderstandings regarding health information when they misinterpret or overvalue the outputs generated by AI. As patients grow more dependent on AI tools, their distrust of physicians can also increase. This is especially true when there is a conflict between the physician's recommendations and the suggestions inferred from

AI. In essence, AI can create a false sense of autonomy, where patients feel empowered to choose from algorithmically predefined options without fully understanding them or engaging in meaningful discussions.

3.3 The Interpretive Model (InT-M) Under AI

The Interpretive Model aims to help patients clarify their values and select treatments aligned with those values, assuming that "the patient's values are not necessarily fixed and known to the patient" [3, p. 2222]. Here, the physician serves as a translator and counselor. But what happens when AI enters this interpretive space? AI might support value clarification—e.g., via pattern recognition in patient narratives—but its presence raises deep questions. Can algorithms grasp the nuance of human values? Should physicians defer to AI as a kind of counselor of counselors?

These questions point to an epistemic risk: the displacement of empathetic, dialogical care by automated inference. As AI appears to successfully "interpret" patients, physicians may gradually disengage from value interpretation, delegating that task to machines. Ahmad et al. show that increased AI use correlates with reduced human motivation, noting that AI "usage and dependency… automatically limit the human brain's thinking capacity," and even "increase human laziness by 68%" [8, p. 4]. In other words, at some point, in the application of AI in medical activities, we would not even know (or care enough) that physician and patient thinking have been undermined through algorithmic shaping of self-perception and professionalism.

3.4 The Deliberative Model (DeL-M) Under AI

The Del-M positions the physician as a guide who helps patients to "elucidate the types of values embodied in the available options" [3, p. 2222]. Considering the Del-M value-laden view, AI could enrich this model by offering structured decision-support systems or value-sensitive scenario modeling. However, the risk is that physicians, overwhelmed by complexity or seduced by efficiency, may overly rely on algorithmically ranked "value hierarchies," unconsciously becoming conduits of AI guidance rather than genuine deliberators. Similarly, patients may benefit from AI-driven scenario modeling. Nonetheless, in turn, they may feel subtly pressured by the apparent neutrality and authority of AI recommendations, perceiving the physician not as a conversational partner but as an amplifier of machine outputs. Therefore, the presence of AI may dangerously erode the co-construction of values that are central to Del-M. Without careful epistemic, educational, and ethical safeguards, the Del-M could be hollowed out—leaving behind the appearance of autonomy without its substance.

4 The Most Suitable Model in an AI-Driven World: The *Philosophically Pluralistic Model*

In this section, my goal is to outline a model of the physician-patient relationship that might best accommodate the ethical, scientific, philosophical, and practical challenges introduced by AI, given the current and near-future landscape in medicine. This model must answer two different situations: one where the physician could have longer and more qualitative time with patients, and another where the physician needs to answer emergency-paced clinical settings—such as, but not only, crowded hospitals, life threatening cases, and high-risk scenarios.

To this aim, while reformulating and adapting the key question in Emanuel and Emanuel's paper [3] to our time, one should ask: how to picture the ideal physician-patient interaction in an AI-driven world (i.e., a world where AI permeates everybody's life and activities in a variety of ways)? To provide an answer to this question, I argue that none of the models above suitably frames the characteristics, issues, and needs of contemporary physician-patient epistemologies, conflicts, and practices. Nonetheless, among the four models (without AI), there are beneficial aspects to the formulation of a new model (that includes AI).

The Pat-M brings one positive advantage besides all its negative aspects (including full responsibility).[1] The positive side of the Pat-M's approach is its capacity to answer the needs of many types of emergency-paced clinical settings. This quality of Pat-M may not only be lifesaving, but it also helps patients to see their doctor and go home faster. One should not dismiss this quality or positive side of the Pat-M. More than that, physicians and medical students in the final years of study apply this quality when they deem it necessary and most efficient.[2]

This highlighted quality of Pat-M (i.e., solving problems in emergency-paced settings) is precisely the weakness of the InF-M, InT-M, and Del-M. For instance, the InF-M wants to ensure individually tailored care—which is even more possible with AI. However, its intent becomes unrealistic since it easily overwhelms patients with complex and overabundant information—which is impractical in emergency-paced scenarios. The InT-M encourages introspection and dialogue, and the DeL-M encourages trust and co-constructed decisions. However, at best, all these models are better suited for not emergency-paced scenarios. In emergency-paced scenarios, all the risks they bring to the table, plus the ones brought by AI, would grow dramatically in quality and quantity.

[1] Full responsibility falling on the physician is taken as a negative effect of the Pat-M, since it can be argued that no physician would want to be fully (exclusively) responsible for whatever outcomes of a medical situation. In some cases, the choices available are simply not optimistic, and whenever possible, it would be good for physicians and for patients to share the choices and consequences.

[2] Sometimes, they are forced to fully adopt the Pat-M. For instance, there are some cases in Brazil where physicians are pressured to spend only 15 min with the patient. In cases like these, the use of AI by physicians will definitely increase the risk that, in practice, AI recommendations get no medical supervision at all.

Furthermore, even in a calm clinical environment, the dialogue and trust between physicians and patients, facilitated by AI, can only thrive when both parties share a certain level of understanding and knowledge on various topics. Given these considerations, I propose a new framework called the *Philosophically Pluralistic Model* (PhiPlur).

The PhiPlur model is based on Feyerabend's philosophy [9], which contends that there is no single form of rationality or authority in science, including in medicine. This means that neither the physician, nor the patient, nor any AI can claim ultimate authority. Thus, the best clinical relationship recognizes that medicine allows diversified, conflicting rationalities and solutions to diseases to coexist within decision-making about theories, methods, and medical responses. In this sense, when we are making a decision, many things come into play, like traditional epistemic values (a plurality of theories, methods, and techniques) and also medical procedures, and ethical and human values.

According to Feyerabend, humanitarian values, such as freedom of thought and human well-being, are cornerstones of what to pursue and avoid. He indicated that his threshold for intervening in other cultures is reached when individuals begin to suffer [10]. This means that suffering defines the moment when intervention becomes necessary. Consequently, the goal of preventing human suffering takes priority over the pursuit of healing. Previous work on Feyerabend's pluralism and education [11] highlights that pluralism alone is folly intellectualism. It is essential to incorporate contextual and practical elements, including the values, desires, and possibilities of the patient. Therefore, the here outlined model of the physician-patient relationship and AI strives to take into account these important factors.

4.1 The Core Premise and Structure of the PhiPlur Model

The PhiPlur model emphasizes the importance of open and constructive communication between physicians and patients facilitated by AI systems. This dialogue should embrace a variety of epistemological traditions, including biomedical, individual, cultural, and algorithmic perspectives. The goal is to enhance the patient's epistemic and existential proficiency, allowing them to navigate, challenge, and select among different worldviews in the context of their care [11, 12]. Medical science is not a single, uniform tradition; rather, it is diverse and pluralistic. It arrives at its conclusions through a wide range of theories and perspectives. Often, these conclusions are partial, provisional, and subject to further testing that may take years to conduct within the general population. In some instances, these findings are proven false, while others evolve into new understandings. This is particularly true for many pharmaceutical drugs.

Second, physicians should help patients critically evaluate the dominant narratives that come from artificial intelligence (e.g., prognostic AI). This reflection should assist both physicians and patients in developing their skills. Third, treatment plans ought to arise from an open and respectful dialogue between physicians and patients—using a Feyerabendian perspective, this means fostering a free exchange of ideas [9]. AI should be utilized as a tool to create alternatives rather than as an objective source of decisions. Therefore, AI must serve as an auxiliary resource, not as the primary one. Fourth, the PhiPlur application must address realistic scenarios that require a dialogue between medical theory and practice, namely the emergency-paced scenarios (red cases) and the not emergency-paced scenarios (green cases).

In the *red cases*, there is a need for immediate action, which suggests that medical handbook answers (*status quo*) have proven to be effective in stopping/reducing people's suffering. This is a decision taken beforehand based on epistemic pluralism, human knowledge, and responsibility. So, healing could be an option when within the *status quo* horizon.

In *green cases*, the philosophically pluralistic model makes more use of alternative theories, methods, epistemologies—combined or not with science—and worldviews. The main goal is not necessarily to heal the patient but, again, to stop/minimize people's suffering. In some cases, such as in some cancers—which will be explored further on—where the chances of healing could be very low, this model flourishes.

Generally speaking, this model embodies the patient as the *decision owner*, even when assisted by AI. The patient is seen as an agent with *unique experiential knowledge*—bodily, cultural, emotional, and spiritual—guided to articulate, adapt, and refine their own meaning-making frameworks in the face of illness. On the other hand, physicians must be treated by patients as scientific experts (although not in an authoritarian way), someone with high proficiency in the game of medical science. Due to this proficiency in medicine and to the pluralistic nature of medicine and science, physicians should act as *pluralist decision facilitators* who are responsible for presenting, creating, and debating alternatives. Considering that physicians have limited time, AI would be helpful for searching for new and ongoing research, creating and exploring possibilities, and learning more about specific diseases, cultures, and values. In this sense, even AI's analysis of a patient's survival for each case must be debated, deeply explored, and not blindly followed.

4.2 The Emergency-paced and Low-Acuity Clinical Setting: Red Cases

In *emergency-paced settings*, here identified as *red cases*, healthcare elements such as time, resources, and decision-making capacity are severely constrained—typically due to high patient volume, high cognitive load on physicians, time pressure, medical and institutional protocols to follow guidelines, or health emergencies. These settings prioritize efficiency and handbook standardization over philosophical reflection.

Similarly, there are also *low-acuity clinical settings*, which differ from emergency-paced settings with respect to urgency and emergency but still share similar constraints on time, personalization, and deliberation. These low-acuity settings, such as cases of influenza, generally do not carry high risks to people's lives.

So, the patient and the physician may not even need to use AI. With enough information about the patient, AI could be seen as the physician assistant. Thus, while low-acuity cases are not like emergency-paced cases, they are still part of the same kind of settings of pressures and responses. Non-exhaustive examples of red case settings are acute stroke management and polytrauma. Examples of low-acuity settings are colds, flu, and bacterial tonsillitis.

In *red cases*, the adoption of quick response helps to reduce human suffering and better apply human and medical resources. In these cases, AI systems may be used to accelerate decisions and help with rapid triage (e.g., symptom checking and registration, antibiotic stewardship alerts, and summarizing patient records). However, as previously mentioned, applying PhiPlur model for these settings still reduces resistance to dominant narratives—like medical and algorithmic authority. Consequently, patients and physicians become more passive and dependent on AI, whereas sociocultural contexts are generalized from regional or national patterns extracted from AI databases or trainings.

For these reasons, the pluralistic model (PhiPlur) adopts part of the response framework provided by the PaT-M. In the case of serious health issues, it is hard to see how even the public would want to alter this approach. For instance, if you visit a hospital with bacterial tonsillitis, the last thing you would want is to wait longer for an effective treatment protocol. This would typically include antibiotics like benzylpenicillin benzathine, a well-known form of penicillin in Brazil. Furthermore, it is unlikely that someone suffering from bacterial tonsillitis would prioritize discussions about cultural values over healing/suffering reduction.

If a patient has previously expressed or documented a wish to limit certain types of procedures or medical treatments, such as blood transfusions or specific medications, AI can quickly alert the physician and explore alternative options. If alternative solutions are not available, the patient will be informed and given the opportunity to decide how to proceed. However, if patients are unconscious, there is no prior information about their preferences, and no one is available to make decisions on their behalf, the medical protocol should be followed.

4.3 The Not Emergency-paced Scenarios. A Real Clinical Case of Breast Cancer: Green Cases

In the so-called not emergency-paced scenarios or green cases, the PhiPlurmodel can reveal its true potential. In the PhiPlurmodel, the responsibility is shared, alternative solutions are welcomed, and exploration of what AI could bring to the table in terms of medical insights, theories, procedures, and alignment of values are encouraged, but not blindly trusted.

All of that is with the aim of addressing patient suffering, not necessarily healing. Take, for instance, the following real case published by Neelima Vidula and colleagues [13]. This case presents a 45-year-old woman diagnosed with metastatic triple-negative breast cancer (TNBC), a subtype known for its aggressiveness, limited targeted therapy options, and poorer prognosis. Her diagnostic journey and treatment trajectory illustrate the clinical, emotional, and ethical complexities central to the physician-patient relationship—now intensified by the introduction of AI in decision-making. According to the authors, the patient was good until "she noticed a mass in the right breast. She was evaluated in the oncology clinic of another hospital, and mammography and ultrasonography of the right breast confirmed the presence of a mass". These evaluations required further testing to confirm the diagnosis, so pathological examination "confirmed the diagnosis of invasive ductal carcinoma." [13, p. 2011].

Additionally, the cancer showed metastatic progression combined with neurological involvement. According to the authors, after some time, the patient developed neurological symptoms, including weakness and numbness in her right arm, where "treatment with dexamethasone and levetiracetam was started", and they did address human suffering, so even from the PhiPlur it was a value alignment treatment. Nevertheless, neuroimaging revealed a 3.5 cm partially cystic mass in the right precentral gyrus of the brain, with surrounding edema and mass effect, so she underwent a "right frontal craniotomy [...], and the mass was resected." [13, pp. 2011–2012].

At this stage, while applying the PhiPlur, doctors should avoid pressuring the patient to make a hasty decision regarding surgery. However, time is of the essence and can provide an opportunity to revisit and reassess the patient's values. This is why, according to PhiPlur, both the physician and AI can assist in this reevaluation process. After all, is the patient willing to undergo those surgeries or prefer to attenuate the pain? What are her values about quantity and quality of life? Questions like these revealed the complexity of clinical judgment, the biological limitations of targeted therapies in TNBC, and the emotional weight of recurrent disease involving the central nervous system.

4.4 The Benefits of the Philosophically Pluralistic Model (PhiPlur) Balanced with AI

From an AI-physician-patient relationship standpoint, such cases demand a delicate balance between providing hope, delivering bad news, and navigating value-laden decisions about invasive procedures, uncertain treatment efficacy, and long-term quality of life. Furthermore, in an AI-driven context, this scenario also raises pressing questions about how AI can improve anything. I contend that, with a pluralistic stance and patient suffering in mind, AI could aid in finding experimental therapies, exploring research, and analyzing patient exams. In medical situations fraught with uncertainties, AI can aid physicians and patients by surfacing lesser-known therapies or simulating outcomes based on ethically filtered variables. It could assist in retrieving case-based reasoning from rare similar conditions without establishing them as objective truths. That being said, and based on the ideas proposed in Sects. 3.3 (about whether AI can grasp human values) and 4.1 (about how AI can support patients and doctors in reassessing values), it is important to say that in PhiPlur, AI helps inform or surface possibilities, but never replaces the human dialogical interpretation of values.

The ways AI could help mediate physician-patient relationships go beyond medical advice. AI could even help patients to prepare for their final days and value realignments, not as human replacement but as a tool. AI can be useful in aiding us to complete projects and support non-scientific beliefs (such as religious ones), as well as helping patients to learn from others who have had similar experiences. Thus, in cases of palliative care settings, AI can identify best practices across cultural or spiritual contexts—providing the patients with options that resonate with their worldview and reinforcing the idea that healing is not only biomedical but existential and symbolic as well. Paul Feyerabend is one example of a somewhat similar case. As he was closer to death, he continued to write his autobiography, something that certainly reduced his suffering—since it changed his focus—provided purpose, and gave him more dignity in dying.

Institutionally, PhiPlur model encourages organizational design where AI is integrated with safeguards for pluralism and human well-being. The PhiPlur, when moderately and wisely coupled with AI, fosters a unique clinical ecosystem where pluralistic rationalities, technology, and compassionate care converge—amplifying human potential to grow rather than reduce or be replaced. Unlike models that either uncritically centralize AI—as an authority—or wholly reject its participation, PhiPlur situates AI as a dialogical *tool*—a cognitive prosthesis— never an agent.

5 Conclusion

In an era where AI increasingly mediates our most intimate and complex human interactions, including healthcare, the physician-patient relationship must evolve beyond the limits of traditional models. Through a critical engagement with Emanuel and Emanuel's four models, I have shown that each one, while historically insightful, lacks the conceptual flexibility to address the epistemic and ethical pressures introduced by AI. The deterministic pull of algorithmic authority risks undermining both the physician's clinical judgment and the patient's autonomy—not merely in decision-making but in the construction of meaning, values, and lived experience under illness. To address this issue, I propose the Philosophically Pluralistic Model (PhiPlur) as an alternative to the four traditional models. This model encourages physicians to view AI as a helpful partner in patient care, but not as an authoritative decision-maker. The PhiPlur offers three key takeaways: First, AI should be seen as a tool rather than a decision-maker. This means that AI outputs should be considered one of many inputs in a larger decision-making process. Second, especially in green cases, physicians must prioritize patient values, aiming first to minimize suffering (when healing is not realistic or the cost-balance of healing misaligned with patient values). Third, AI's applications vary according to the settings (red or green cases), and physicians and hospitals may need to adapt as soon as possible. By embracing a pluralistic and flexible approach, the physician-patient relationship can maintain its ethical depth in an increasingly AI-mediated clinical environment. More philosophically, while the PhiPlur resists the consolidation of authority—whether from physicians or AI—and instead affirms care as a space of dialogue, situated reasoning, and mutual responsibility. If the future of medicine is to be both technologically advanced and ethically sound, then the relationship between physicians and patients must embody a shared commitment to meaning and value, fostering trust, preserving human dignity, and focusing on the ultimate goal of alleviating suffering—which could or could not be realized with healing.

Competing Interests The author has no conflicts of interest to declare that are relevant to the content of this chapter.

References

1. Suleyman M (2023) The coming wave: technology, power, and the twenty-first century's greatest dilemma. Crown, New York
2. Yokoi R, Eguchi Y, Fujita T, Nakayachi K (2021) Artificial intelligence is trusted less than a doctor in medical treatment decisions: influence of perceived care and value similarity. Int J Human-Comput Interact 37:981–990. https://doi.org/10.1080/10447318.2020.1861763
3. Emanuel EJ, Emanuel LL (1992) Four models of the physician-patient relationship. JAMA 267:2221–2226
4. Nuland SB (2004) The doctors' plague: germs, childbed fever, and the strange story of Ignac Semmelweis (great discoveries). WW Norton & Company
5. Nowotny H (2021) In AI we trust: power, illusion and control of predictive algorithms. Polity, Cambridge, UK
6. Cafardo R (2025) O que fazer com 50 mil medicos formados em cursos mal avaliados pelo MEC https://www.estadao.com.br/educacao/renata-cafardo/o-que-fazer-com-50-mil-medicos-formados-em-cursos-mal-avaliados-pelo-mec/?srsltid=AfmBOooOQWvV1h1ubw4rzyvnBtAT3kXBzf8oXS2UbtzgCP537Z8af3SU. Accessed on 06 May 2025
7. Susser D (2019) Invisible influence: artificial intelligence and the ethics of adaptive choice architectures. In: Proceedings of the 2019 AAAI/ACM conference on AI, ethics, and society, pp 403–408
8. Ahmad SF, Han H, Alam MM, Rehmat MK, Irshad M, Arraño-Muñoz M, Ariza-Montes A (2023) Impact of artificial intelligence on human loss in decision making, laziness and safety in education. Humanit Soc Sci Commun 10:311. https://doi.org/10.1057/s41599-023-01787-8
9. Feyerabend P (1993) Against method. Verso, New York
10. Feyerabend P (1995) Three interviews with Paul K. Feyerabend. In: Parascandalo R, Hösle V (eds) Telos, vol. 1995, Rome, pp 115–148
11. Oliveira DGS (2024) Rethinking science education: fostering Feyerabend's view of pluralism and proficiency. In: Kenklies K, Engelmann S (eds) Education for a free society Paul Feyerabend and the pedagogy of irritation, Book Series Paedagogica. Peter Lang Publishing Press., New York, NY
12. Oliveira DGS (2021) The cosmological divergent proliferation in Feyerabend's pluralism. Principia 25:421–454. https://doi.org/10.5007/1808-1711.2021.e72764
13. Vidula N, Rodriguez K, Wong AK, Boyraz B (2024) Case 17-2024: a 45-year-old woman with metastatic breast cancer. N Engl J Med 390(21):2011–2022. https://doi.org/10.1056/NEJMcpc2309499

Deivide Garcia da Silva Oliveira is professor of philosophy at the Federal University of Sergipe, Brazil. He completed his PhD in Philosophy and History of Sciences at the Federal University of Bahia-UFBA, Brazil, and held a postdoctoral position at the York University-Canada. His research focuses on scientific pluralism, Feyerabend's philosophy, history and philosophy of science, Galileo's methodology, and Philosophy of Artificial Intelligence.

Big Data and the "Small X". Patient Autonomy in the Age of AI

12

Chantal Marazia, Vasilija Rolfes, and Fabio De Sio

A thinker whom I regard with personal esteem says: "If we call all that an individual man is, has and performs A, than this A arises out of a + x, a embracing all that comes to the man from his outer circumstances: from his country, people, age, etc., while the vanishing little x is his own contribution, the work of his free will. However vanishing small this x may be, it is of infinite value. Morally and humanly considered it alone has value.

Gustav Droysen, Outline of the Principles of History, [1868] 1897.

Abstract

Patient autonomy has become a cornerstone of contemporary medical ethics, reshaping the doctor–patient relationship through the practices of informed consent and shared decision-making. The integration of artificial intelligence (AI) into healthcare introduces new challenges to this principle. While AI promises improved efficiency and personalized care, it also risks entrenching systemic biases, diminishing clinical transparency, and disrupting the relational foundations of care. These concerns are particularly acute for patients from minoritized and culturally diverse backgrounds, whose needs are often excluded from AI training data and system design. Algorithmic opacity, in turn, may threaten the dialogical conditions necessary for meaningful consent, potentially fostering a new form of "double paternalism" where both physicians and patients defer to inscrutable machine authority. This chapter argues that safeguarding autonomy in the age of AI requires more than technical oversight or regulatory safeguards. Drawing on the frameworks of cultural and narrative humility, it proposes a

C. Marazia (✉)
University of Milan, Milan, Italy
e-mail: chantal.marazia@unimi.it

V. Rolfes · F. De Sio
Heinrich-Heine-University Düsseldorf, Düsseldorf, Germany
e-mail: Vasilija.Rolfes@uni-duesseldorf.de

M. Anzalone et al. (eds.), *Reconfiguring Human Autonomy*, Integrated Science 40,
https://doi.org/10.1007/978-3-032-14537-6_12

relational and reflexive approach to care, which foregrounds critical self-awareness and a sustained commitment to co-constructing meaning across difference: algorithmic humility. Only by cultivating these ethical dispositions can clinicians resist reductive models of identity, counterbalance algorithmic influence, and affirm patient autonomy as a dynamic, socially embedded process.

Keywords

Patient autonomy · Artificial intelligence in healthcare · Shared decision-making · Algorithmic bias · Cultural and narrative humility · Algorithmic humility

1 Introduction

Medical ethics has proven resistant to the dominance of any single classical moral theory. Neither the calculative clarity of utilitarianism, nor the rigor of Kantian duty, or Aristotelian virtues have imposed themselves as a steady compass to navigate the value-laden terrain of healthcare. In the face of this moral pluralism, Tom Beauchamp and James Childress' influential framework sidestepped theoretical absolutism in favour of a pragmatic orientation: a flexible, yet structured approach to clinical decision-making as well as to all other levels of the healthcare system, known as "Principlism" [1]. Their model famously centres on four fundamental ethical principles shared across diverse traditions and professional practices: beneficence, non-maleficence, justice, and respect for autonomy.

Historically the youngest of the four principles, patient autonomy has risen to a cornerstone of Western medical ethics, providing a counterbalance to traditional medical paternalism [2]. It encompasses not only a negative dimension—freedom from external coercion and manipulative influence—but also a positive, constructive component: active participation in decision making [1]. In the clinical setting, the principle of autonomy is embodied in the practice of informed consent, which ensures that patients have sufficient information to make voluntary and competent decisions [3]. For the physician, this implies the responsibility of providing careful information tailored to the patient's specific needs, values and goals, and ensuring that patients fully understand their options. Shared decision making at its best requires both accurate information and consideration of personal values [4].

The transition from the older "paternalistic model" to the "autonomy model" has reshaped the doctor–patient relationship, fostering greater transparency, honesty, and a heightened respect for the patient's inherent dignity [5]. This shift has coincided with a broader change in the understanding of autonomy itself. Over the past few decades, the concept has undergone substantial critical revision, notably expanding through the introduction of relational autonomy [6]. This perspective enriches the traditional focus on individual self-determination by emphasising the role of social relationships, cultural contexts, and structural conditions in shaping autonomous agency and enhancing quality of care at the same time.

The accelerating integration of AI into healthcare systems, made at once possible and desirable by the development and integration of unprecedentedly large and persistent datasets (commonly known as Big Data [7]), not only questions the meaning of the traditional, patient-centred concept of autonomy, but also foreshadows new epistemic, normative and overarching ethical concerns. Luciano Floridi and collaborators have devised an analytical tool (cf. 8 and bibliography therein), aimed at helping policymakers and all concerned parties to anticipate and assess these concerns, plotting them at six different levels of abstraction: individual, interpersonal, group, institutional, sectoral and societal.[1] Traditionally, the issue of autonomy is raised at the first two levels, but the large-scale application of algorithms to healthcare and wellbeing widens its import dramatically. This chapter will take the move from the traditional, patient-centred concept of autonomy, and explore its complications, which are increasing in conjunction with its potential.

2 AI and Healthcare

AI has been hailed as one of the most transformative developments in modern medicine [9]. This set of technologies, particularly those used in diagnostic imaging, predictive analytics, clinical decision support systems, and personalised treatment planning, promises to reshape healthcare by improving diagnostic accuracy, optimising treatment strategies, and increasing overall healthcare efficiency [10]. However, these high expectations are tempered by profound ethical concerns [11, 12].

Potential biases and systemic unfairness inherent in AI systems have been the object of sustained attention since the beginning (for systematic reviews cf. [13–16]). The data used to train algorithms are subject to two primary forms of bias: data bias and algorithmic bias [17]. Data bias originates in the processes of data collection and preprocessing, whereas algorithmic bias stems from the design or learning mechanisms embedded within the algorithm itself. A typical indicator of data bias is the underrepresentation of certain minority groups in the dataset, which prevents the algorithm from learning accurate patterns relevant to those populations. As a result, AI systems may perform inadequately for specific groups.

Algorithmic bias encompasses forms such as label bias and cohort bias [18]. Label bias occurs when the training process relies on inconsistent or unreliable labels—often shaped by existing healthcare disparities. This can lead the AI system to make systematically biased decisions based on misleading information. Cohort bias arises when algorithms are trained on easily quantifiable or traditionally studied populations, at the expense of protected groups or more complex populational differences.

[1] Such a schematisation is aimed at making the different impacts evident and treatable by different stakeholders. The Authors make it clear that it does not exclude overlap between the levels.

In the overwhelming flow of literature on the subject, a growing body of scholarship highlights biases related to gender, race, and socioeconomic background (see especially the revealing [19]).

A first source of inequality lies in the composition of AI development teams, which often lack diversity in terms of gender, ethnicity, or professional background. This lack of diversity may critically affect the system's foundational architecture, as developers may not fully account for gender, populational and cultural differences in health-related values, behaviours, and practices. As a result, AI systems may inadvertently prioritise established practices and systematically ignore plausible alternatives. To date, the greatest share of the data used for training AI systems is derived from the demographic landscape and clinical practices typical of Western healthcare systems, which may not reflect the ethnic, cultural and socioeconomic diversity found globally, and thus be not universally applicable. A most palpable example of this is found in dermatology tools, which, predominantly trained on images of lighter skin tones, often underperform in detecting dermatological conditions on darker ones [20]. However, racial and cultural biases apply to virtually all training data, across all medical disciplines. Whereas, in the cases mentioned above, a relatively straightforward answer could be provided by a deliberate expansion of the training data to cover a greater diversity of sources, the more advanced algorithmic models pose a greater challenge, due to their greater processing potential and, consequently, greater "autonomy" of the machine [21, 22], inevitably paralleled by a growing inscrutability of its intrinsic functioning. In the domain of mental health, for instance, natural language processing (NLP) tools are increasingly used to assess patient narratives, extract symptoms, and aid in psychiatric diagnosis [23]. These tools often rely on linguistic analysis of patient narratives. However, language use is deeply shaped by sociocultural expressions of distress, as well as idioms of suffering, and communication norms vary widely. Diversity-insensitive AI systems may misinterpret idiomatic expressions or specific communication based on patients' biography patterns that are integral to accurate mental health assessments. An AI system trained on a limited set of "standard" expressions of depression, for instance, might miss signs in patients from different sociocultural backgrounds. Such biases, once established (see [24]), risk undermining clinical accuracy and entrenching a diversity-myopic model of medicine, thus exacerbating structural inequities and further widening disparities in health outcomes and access to care [25, 26]. From data collection, processing, development, and validation, to their full implementation, AI systems are indeed "predictably unequal" [27].

3 AI and Autonomy

Bias in the application of AI in healthcare can obstruct equitable access to quality care and compromise patient well-being, thereby directly undermining the principles of beneficence, non-maleficence, and justice, at once. Less apparent, yet equally consequential, is its potential to erode patient autonomy in the traditional sense, and to question autonomy in new ways, at several levels of abstraction. As algorithmic systems increasingly participate in the formulation of diagnostic and therapeutic recommendations, the locus of medical authority becomes at best more distributed—and, in some cases, destabilized [28]. There is, of course, a range of perspectives on AI, reflecting diverse attitudes among healthcare providers and patient groups. Individualistic cultures, for instance, may tend to picture AI as external to the self, perceiving it as infringing on uniqueness, autonomy, and privacy, whereas those from collectivist cultures may see it more as an extension of the self, facilitating social harmony, consensus, and protection of privacy [29]. In the clinical context, senior practitioners, shaped by years of empirical judgment and interpretive nuance, tend to regard AI systems with greater caution, suspicious of too formalized decision pathways. In contrast, less experienced practitioners may defer too readily to algorithmic authority, particularly when pressured by institutional demands for efficiency or standardization. These divergent orientations toward AI can impact how medical decisions are framed and negotiated with patients, potentially reducing clinical judgment to a function of technological mediation. In this way, AI may influence not only the structure of decision-making, but also the distribution of epistemic confidence within healthcare teams [30].

Regardless of individual or group attitudes towards AI, the greatest problem raised by the development of this technology is, however, its intrinsic procedural opacity [31]. These tools often generate results through processes defying intelligibility—whether due to proprietary design, computational complexity, or the sheer volume of interacting variables. In such cases, neither physician nor patient may be able to trace the logic of a given recommendation. This epistemic opacity has practical implications: it may inhibit the transparency required for informed consent, generate ambiguity about clinical (as well as legal) responsibility, and erode the normative foundation of the physician–patient dialogue. When AI functions as an inscrutable arbiter of care, both trust and accountability risk becoming abstracted, that is to say, displaced into a system whose authority is opaque and whose fallibility is difficult to assess [32], thus leading to a new type of medical paternalism: double paternalism, i.e. "the establishment of a paternalistic relationship first between the AI and the doctors, and consequently between the doctors and the patients" [24, p. 428] (see also [33]). At the sectoral and societal level of abstraction, this lack of transparency risks undermining public confidence in AI applications generally, beyond the strictly medical domain [34, 35]. Significantly, a number of multi-stakeholder position documents produced in the USA, Canada, UK and EU in the past decade, highlight the critical need for transparency, accountability, and intelligibility of the mechanised decision-making processes. In a critical synthesis of these statements, Floridi et al. have gone so far as raising "explicability" to the level of a fifth principle of biomedical ethics, one specifically "enabling for AI" [36].

In the face of such widespread concerns, it has been objected that the presence of black-box AI in clinical practice does not *inherently* threaten patient autonomy, whether conceived in traditional or relational terms. Rather, autonomy would be shaped by the way AI technologies are employed. If black-box systems ended up fully replacing the clinician, the absence of causal transparency could jeopardize autonomy by disrupting the trust and dialogue central to the patient–physician relationship. However, if accompanied by informed consent, post-hoc explanations, and clear communication, black-box AI could support and even enhance patient autonomy by enabling access to more personalized and effective healthcare options [37].

We contend, however, that the problem reaches deeper than formalised accounting strategies. What is at stake here is not merely technical competence or communicative clarity, but the very conditions under which patient autonomy acquires meaning in the clinical encounter. In order to preserve its meaning, informed consent cannot be reduced to the simple issue of information transfer—the more so when that information originates from systems whose inner workings remain opaque or intractable. In its fullest sense, autonomy presupposes a dialogical space in which the patient's values, self-understanding, and interpretive agency are not only acknowledged but also actively empowered. This entails more than gestural empathy; it demands a sustained attentiveness to both the patient's and the caregiver's lived worlds, and a willingness to inhabit the complexity of illness as it unfolds—with or without artificial intelligence.

4 Humbling Experiences

Our preliminary assessment of AI integration in medicine reveals two pressing concerns. First, AI systems tend to replicate and intensify existing biases, disproportionately affecting minority and marginalised populations, and promoting a model of care that overlooks diversity. Second, the growing reliance on algorithmic tools in clinical settings underscores the need to strengthen the patient–physician relationship and preserve an honestly shared decision making. Notably, both issues are not new: medicine has long grappled with structural inequalities and with the tension between institutional priorities, clinical authority and patient autonomy.

Within the broader discourse on equity and diversity in healthcare, efforts to cope with the heterogeneity and individuality of patient subjects have not been spared.

One such effort, which emerged at the end of the twentieth century, was "cultural competence": a framework to help clinicians work effectively across cultural boundaries, by acquiring knowledge about the beliefs, values, and health practices of different cultural groups. This approach was rooted in the assumption that better understanding of cultural contexts would lead to more accurate diagnoses and more sensitive care delivery [38]. Despite the laudable goal of promoting respect for diversity and individuality, it has faced significant criticism. One of the main objections is that cultural competence tends to reify culture—treating it as a static set of characteristics that can be learned, categorised, and applied to all individuals within a group. Paradoxically, this approach can end up reinforcing stereotypes and essentialist views [39]. By reducing culture to a checklist, cultural competence risks essentializing identities and neglecting intra-group variation [40]. Individuals are shaped by the culture in which they are socialized, an influence that permeates various aspects of life—including social behaviour, the perception of one's physical and mental states, and the very notion of what is deemed "normal." Yet, it is important to recognize that socialization within a given culture is always an ongoing process, allowing individuals a measure of autonomy and flexibility in their actions. Their behaviour is also shaped by prior experiences, personal aspirations, anxieties, and economic circumstances [41]. In light of these concerns, Arthur Kleinman and Peter Benson [38] have proposed a step-by-step anthropological approach to the patient, through productive questions instead of preconceptions and stereotypes on the part of the medical teams. These inquiries, led by medical professionals within a therapeutic context, are designed to bring out the patient's perspective on their illness, any apprehensions they may have regarding treatment, and their personal understanding of the condition. The explicit interest in the patient's perspective preempts any preconceived, culturally determined notions. The "cultural humility"-perspective represents a further step forward in the same vein: a shift away from the mastery of cultural knowledge toward a lifelong commitment to self-evaluation, reflexivity, and the recognition of power imbalances. Coined by Melanie Tervalon and Jann Murray-García in 1998, this approach rejects the idea that clinicians can ever become fully "competent" in someone else's culture [40]. Instead, it encourages professionals to adopt a stance of respectful curiosity, acknowledging patients as the experts of their own lived experience and values. Cultural humility foregrounds relational ethics and emphasizes the ongoing negotiation of meaning between caregiver and patient. The key distinction between the two approaches thus lies in their ethical orientation: Cultural competence is rooted in an epistemological model of expertise, often aligned with institutional training and policy mandates. Cultural humility,[2] by contrast, is grounded in moral reflexivity, fostering critical awareness of one's own cultural lens and institutional privileges. While some scholars advocate for integrating the two approaches [42]—retaining the pragmatic benefits of cultural knowledge while embedding it within a humble and critically aware posture—others caution that any such integration must resist reverting to a model of cultural "expertise". Cultural humility is ultimately less about what providers know and more about how they relate to others in the face of uncertainty, difference, and complexity.

This ethical recalibration—from knowing about culture to relating across differences—finds a resonant parallel in the emerging discourse on narrative humility. Just as cultural humility resists the reduction of cultural identity to static knowledge, narrative humility challenges the traditional paradigms of medical authority and epistemic certainty. Coined by Sayantani DasGupta within the framework of

[2] While often used interchangeably, cultural sensitivity and cultural humility represent distinct ethical orientations. The former involves an awareness of and respect for cultural differences. Applied to the medical context, it encourages healthcare professionals to acquire knowledge about other cultures and apply that understanding in clinical interactions. However, this approach can risk becoming static or superficial if it relies on generalized cultural profiles rather than engaging with the patient as a unique individual. Cultural humility, by contrast, emphasizes an ongoing process of self-reflection, learning, and power-sharing. It requires practitioners to recognise the limitations of their own cultural perspective, to listen actively to patients' lived experiences, and to remain open to revising assumptions.

narrative medicine [43], the concept underscores the ethical necessity of acknowledging the limits of one's understanding when encountering patients' stories. By embracing the ambiguity and moral depth of illness narratives, clinicians can foster a richer, more relational kind of knowledge—one that attends not only to pathology, but to suffering, identity, and the lived texture of human experience. Like the "culturally humble" approach, narrative humility emphasizes an ongoing stance of openness, reflexivity, and ethical responsiveness. In addition, narrative humility places a greater emphasis on a dialogical approach to care—one in which meaning is co-constructed and mutual vulnerability forms the basis for genuine connection. Narrative humility thus requires more than attentive listening; it demands a readiness to be changed by what one hears. Only through such relational work can decisions emerge that are both ethically sound and existentially resonant. Communication in which the physician listens, and the patient's narrative is valued, can increasingly allow their individual realities of life to converge [44, 45]. In this sense, the patient has the opportunity to give space to interpretation of the illness, own life circumstances, and already existing experiences with the illness. Such openness is a foundational prerequisite that helps prevent physicians' assumptions from being decisively shaped by factors such as "age, disease or disability, creed, ethnic origin, gender, nationality, political affiliation, race, sexual orientation, social standing, or any other factor" [46]. This stance is particularly vital in multicultural or marginalized settings, where it helps prevent the erasure of difference and affirms the patient's voice as central to the therapeutic process. Additionally, narrative humility offers a compelling counterpoint to the technocratic and efficiency-driven imperatives of contemporary health management. Narrative humility, in fact, calls for emotional receptivity and epistemic humility, urging clinicians to reflect critically on their own assumptions, positionalities, and interpretive frameworks, and to recognize how these influence the clinical encounter.

5 Algorithmic Humility

A further humbling experience comes from the acknowledgement that neither the attention towards life-histories, nor the dream of a Big Data-driven science are at all novel. At different junctures in the development of his *General Psychopathology*, the German Philosopher Karl Jaspers had in fact already voiced both. "Psychiatry", he remarked already in 1923,[3] "urgently requires a register [*Archiv*] of case-histories which are biographical in nature and worked over thoroughly. The usual chaotic notes are not at all satisfactory. Mechanical case-histories, written according to stupid rules of 'objectivity' and reproducing every sentence noted in the case-record, are profitless. We must have definite points of view and an eye for significant facts" [47, p. 572]. It is fair to point out that, in making this point, Jaspers was promoting

[3] We cite here the 1963 English translation of the fourth edition (1946) of Jaspers' *General Psychopathology*, although the paragraph in question originally appeared in the second edition, published in 1923.

an epistemic and epistemological agenda: improving diagnostic power, and consolidating psychiatry as a science [48, 49]. The critical ethical issue of patient autonomy was outside of his horizon. Yet, one of the theoretical underpinnings of his biographical approach, Gustav Droysen's methodology of history [50], had already highlighted the inextricable connection between the epistemological and moral domains: "However vanishing small [the] x may be, it is of infinite value. Morally and humanly considered it alone has value" [51, p. 78].

Building on different but convergent sources, Silvia Pierosara has convincingly argued for the interdependence of narrative humility and autonomy in the context of care: "Where autonomy and humility intersect," she writes, "transformation and consolation—segments of a story with at least two authors—stand side by side. If their normative core resides in the duty to not settle for surrogates of meaning, not fabricating oversimplified ones when meaning cannot be seen but is constructed together, their ethical core is to be found in the reciprocal capacity to decentralise for the benefit of the stories to come" [52]. Pierosara explicitly builds on the concept of narrative humility. As Antonio Casado da Rocha argues, autonomy involves two interrelated dimensions: "informational autonomy (the capacity to manage their health information in the present moment) and narrative autonomy (the capacity of patients to develop a shared story linking their past and future)" [53, p. 207]. The challenge—in this story like in Droysen's conception of history—is to uncover and give voice to the *little x*: in the singular, at the individual level of abstraction; in the plural "*little xs*", at the interpersonal level.

Recognising and responding to these "little xs" is essential not only for upholding the autonomy of individual patients, but also for enabling the autonomy of caregivers, understood as the capacity to act responsively and ethically in light of someone else's story. Shared decision-making, at its core, requires that both parties—patient and caregiver—are able to bring their positions and values into a dialogical space where mutual understanding can unfold. Autonomy, in this view, is not simply individual independence, but rather relational co-authorship. AI can thus play a constructive role in enabling the ethical dynamics at the heart of shared clinical decision-making. With a measure of warranted optimism, AI may support the clinical encounter in two interrelated ways: first, as a tool, and second, as an experiential horizon. As a tool, trivially enough, AI can assist by automating routine and low-complexity tasks—such as drafting patient records [54]—thus freeing up time and cognitive resources to be reallocated toward the more meaningful, person-centered aspects of care. This is a practical gain, but also an ethical one, insofar as it potentially restores focus on the relational and dialogical dimensions of the clinical encounter.

Yet AI's most transformative potential arguably lies in its role as an experiential horizon—one that invites a reconfiguration of traditional therapeutic hierarchies. The broader integration of AI into medicine continues a long arc of rationalisation and standardisation, echoing earlier developments such as clinical protocols and evidence-based meta-analyses. However, unlike these predecessors, AI systems introduce a qualitatively distinct challenge: they generate outputs whose epistemic grounding is to some extent inaccessible not just to patients, but even to clinicians themselves. This epistemic opacity cannot be managed by simply relegating AI tools to a supplementary or advisory role. Rather, it compels us to articulate a collective understanding of the epistemic status of algorithmic outputs, as well as the norms governing their influence on decision-making. Precisely because the functioning of algorithmic systems resists full transparency, even in principle, their presence in clinical practice demands an interpretive effort shared by both physician and patient. In this sense, algorithmic opacity becomes not a limitation, but a provocation: it can catalyse co-reflection and co-deliberation, encouraging both parties to rethink their roles and to negotiate authority in new ways. Such dynamics, far from undermining patient autonomy, may actually deepen it, pushing it beyond the minimalist version of informed consent toward a richer ideal of participatory agency.

In the light of these considerations, the notion of narrative humility can be taken further. Far from being a soft skill or secondary virtue, narrative humility is a critical disposition: it requires clinicians to interrogate their own assumptions and interpretive frames, to remain emotionally and cognitively receptive, and to engage seriously with the experiences and values of others. Taken seriously, this stance must also encompass the patients' and the clinicians' own attitudes toward AI, whether these lean toward scepticism or overreliance.

When we accept that the logic underpinning algorithmic tools exceeds the clinician's domain of exclusive expertise, we are led to what may be called algorithmic humility.[4] This is not merely a matter of ethical benevolence or cautious restraint. It is a substantive acknowledgment that algorithmic systems generate a surplus of meaning and operational logic irreducible to either human subject's understanding. Recognising this excess entails a shift from control to cohabitation—from a model of mastery to one of reflective partnership. And it is in this shift that a renewed vision of autonomy and responsibility may begin to take shape. Algorithmic humility, as such, is not a simply a competence, acquired once and for all, but a veritable epistemic virtue [55], caused by an epistemic necessity—one that must be continually cultivated, renewed, and tested within the specificities of each clinical encounter and technological scenario.

6 Core Messages

- Patient autonomy must be reinterpreted while considering AI integration, emphasizing relational, culturally sensitive, and narratively grounded approaches to consent and shared decision-making.

[4] We use the expression "algorithmic humility" differently from how it was defined at the Fourth Machine Learning for Health (ML4H) symposium (December 15–16, 2024), as "an approach to AI system design" that "seeks to develop AI systems that not only address human cognitive biases but also resist becoming unquestioned sources of authority themselves" [56]. We intend "algorithmic humility" as an individual epistemic virtue.

- Protecting patient autonomy demands more than technical disclosure—it requires relational and narrative humility that affirms patients as culturally situated individuals and co-constructors of meaning in the therapeutic encounter.
- Rather than striving for full explainability, the ethical integration of AI in medicine requires *algorithmic humility*—a reflective stance that embraces uncertainty, resists reductionism, and prioritizes ongoing ethical engagement with the relational and cultural complexity of clinical care.

Competing Interests The authors have no conflicts of interest to declare that are relevant to the content of this chapter.

References

1. Beauchamp TL, Childress JF (1979) Principles of biomedical ethics. Oxford University Press, Oxford
2. Will JF (2011) A brief historical and theoretical perspective on patient autonomy and medical decision making: part II: the autonomy model. Chest 139(6):1491–1497. https://doi.org/10.1378/chest.11-0516
3. Berg JW, Appelbaum PS, Lidz CW, Parker LS (2001) Informed consent: legal theory and clinical practice. Oxford University Press, Oxford/New York
4. Stiggelbout AM, Van der Weijden T, De Wit MPT, Frosch D, Légaré F, Montori VM, Trevena L, Elwyn G (2012) Shared decision making: really putting patients at the centre of healthcare. BMJ 344:e256. https://doi.org/10.1136/bmj.e256
5. Pellegrino ED (1990) The relationship of autonomy and integrity in medical ethics. Bull Pan Am Health Organ 24(4):361–371
6. Dive L, Newson AJ (2018) Reconceptualizing autonomy for bioethics. Kennedy Inst Ethics J 28(2):171–203. https://doi.org/10.1353/ken.2018.0013
7. Aicardi C, Reinsborough M, Rose N (2018) The integrated ethics and society programme of the human brain project: reflecting on an ongoing experience. J Responsible Innov 5(1):13–37. https://doi.org/10.1080/23299460.2017.1331101
8. Morley J, Floridi L (2025) The ethics of AI in healthcare: an updated mapping review. In Altman MC, Schwan D (Eds), Ethics and medical technology: essays on artificial intelligence, enhancement, privacy, and justice, pp 29–57
9. Noorbakhsh-Sabet N, Zand R, Zhang Y, Abedi V (2019) Artificial intelligence transforms the future of health care. Am J Med 132(7):795–801. https://doi.org/10.1016/j.amjmed.2019.01.017
10. Topol EJ (2019) Deep medicine: how artificial intelligence can make healthcare human again. Basic Books, New York
11. World Health Organization (WHO) (2018) Ethical issues in the use of AI in health care. WHO, Geneva
12. Char DS, Shah NH, Magnus D (2018) Implementing machine learning in health care—addressing ethical challenges. N Engl J Med 378(11):981–983. https://doi.org/10.1056/NEJMp1714229
13. Ferrara E (2023) Fairness and bias in artificial intelligence: a brief survey of sources, impacts, and mitigation strategies. Sci 6(1):3. https://doi.org/10.3390/sci6010003
14. Pagano TP, Loureiro RB, Lisboa FV, Peixoto RM, Guimarães GA, Cruz GO et al (2023) Bias and unfairness in machine learning models: a systematic review on datasets, tools, fairness metrics, and identification and mitigation methods. Big Data Cogn Comput 7(1):15. https://doi.org/10.3390/bdcc7010015

15. Shams RA, Zowghi D, Bano M (2025) AI and the quest for diversity and inclusion: a systematic literature review. AI Ethics 5:411–438. https://doi.org/10.1007/s43681-023-00362-w
16. Pressman SM, Borna S, Gomez-Cabello CA, Haider SA, Haider C, Forte AJ (2024) AI and ethics: a systematic review of the ethical considerations of large language model use in surgery research. Healthcare 12(8):825. https://doi.org/10.3390/healthcare12080825
17. Ueda D, Kakinuma T, Fujita S et al (2024) Fairness of artificial intelligence in healthcare: review and recommendations. Jpn J Radiol 42:3–15. https://doi.org/10.1007/s11604-023-01474-3
18. Rajkomar A, Hardt M, Howell MD, Corrado G, Chin MH (2018) Ensuring fairness in machine learning to advance health equity. Ann Intern Med 169(12):866–872. https://doi.org/10.7326/M18-1990
19. Buolamwini J, Gebru T (2018) Gender shades: intersectional accuracy disparities in commercial gender classification. In: Proceedings of the 1st conference on fairness, accountability and transparency. ACM, New York, pp 77–91. https://doi.org/10.1145/3287560.3287593
20. Adamson AS, Smith A (2018) Machine learning and health care disparities in dermatology. JAMA Dermatol 154(11):1247–1248. https://doi.org/10.1001/jamadermatol.2018.2348
21. Pierosara S (2024) Narrative autonomy and artificial storytelling. AI Soc 39(4):1785–1794. https://doi.org/10.1007/s00146-022-01595-9
22. Haltaufderheide J, Ranisch R (2024) The ethics of ChatGPT in medicine and healthcare: a systematic review on Large Language Models (LLMs). NPJ Digit Med 7(1):183. https://doi.org/10.1038/s41746-024-01157-x
23. Malgaroli M, Hull TD, Zech JM, Althoff T (2023) Natural language processing for mental health interventions: a systematic review and research framework. Transl Psychiatry 13:309. https://doi.org/10.1038/s41398-023-02592-2
24. Panch T, Mattie H, Celi LA (2019) The "inconvenient truth" about AI in healthcare. NPJ Digit Med 2(1):77. https://doi.org/10.1038/s41746-019-0155-4
25. Obermeyer Z, Powers BW, Vogeli C, Mullainathan S (2019) Dissecting racial bias in an algorithm used to manage the health of populations. Science 366(6464):447–453. https://doi.org/10.1126/science.aax2342
26. Haider SA, Borna S, Gomez Cabello CA et al (2024) The algorithmic divide: a systematic review on AI-driven racial disparities in healthcare. J Racial Ethn Health Disparities. https://doi.org/10.1007/s40615-024-02237-0
27. Fuster A, Goldsmith-Pinkham P, Ramadorai T, Walther A (2022) Predictably unequal? The effects of machine learning on credit markets. J Finance 77(1):5–47. https://doi.org/10.1111/jofi.13090
28. Lorenzini G, Arbelaez Ossa L, Shaw DM, Elger BS (2023) Artificial intelligence and the doctor–patient relationship: expanding the paradigm of shared decision making. Bioethics 37(5):424–429. https://doi.org/10.1111/bioe.13158
29. Barnes AJ, Zhang Y, Valenzuela A (2024) AI and culture: culturally dependent responses to AI systems. Curr Opin Psychol:101838. https://doi.org/10.1016/j.copsyc.2024.101838
30. Gaube S, Suresh H, Raue M et al (2021) Do as AI say: susceptibility in deployment of clinical decision-aids. NPJ Digit Med 4:31. https://doi.org/10.1038/s41746-021-00385-9
31. Price WN (2018) Big data and black-box medical algorithms. Sci Transl Med 10(471):eaat2864. https://doi.org/10.1126/scitranslmed.aat2864
32. Rolfes V, Bittner U, Gerhards H, Krüssel JS, Fehm T, Ranisch R, Fangerau H (2023) Artificial intelligence in reproductive medicine – an ethical perspective. Geburtshilfe Frauenheilkd 83(1):106–115. https://doi.org/10.1055/a-1866-2792
33. Xu H, Shuttleworth KMJ (2024) Medical artificial intelligence and the black box problem: a view based on the ethical principle of "do no harm". Intell Med 4(1):52–57. https://doi.org/10.1016/j.imed.2023.08.001
34. Rudin C (2019) Stop explaining black box machine learning models for high stakes decisions and use interpretable models instead. Nat Mach Intell 1:206–215. https://doi.org/10.1038/s42256-019-0048-x
35. von Eschenbach WJ (2021) Transparency and the black box problem: why we do not trust AI. Philos Technol 34(4):1607–1622. https://doi.org/10.1007/s13347-021-00477-0

36. Floridi L, Cowls J, Beltrametti M et al (2018) AI4People—an ethical framework for a good AI society: opportunities, risks, principles, and recommendations. Minds Mach 28:689–707. https://doi.org/10.1007/s11023-018-9482-5
37. Prince S, Lim JE (2025) Black-box AI and patient autonomy. Minds Mach 35(2):1–19. https://doi.org/10.1007/s11023-025-09729-w
38. Kleinman A, Benson P (2006) Anthropology in the clinic: the problem of cultural competency and how to fix it. PLoS Med 3(10):e294. https://doi.org/10.1371/journal.pmed.0030294
39. Carillo JE, Green AR, Betancourt JR (1999) Cross-cultural primary care: a patient-based approach. Ann Intern Med 130(10):829–834. https://doi.org/10.7326/0003-4819-130-10-199905180-00017
40. Tervalon M, Murray-García J (1998) Cultural humility versus cultural competence: a critical distinction in defining physician training outcomes in multicultural education. J Health Care Poor Underserved 9(2):117–125. https://doi.org/10.1353/hpu.2010.0233
41. Verwey M (2003) Hat die Odyssee Odysseus krank gemacht? Migration, integration und Gesundheit. In: Lux T (ed) Kulturelle Dimensionen der Medizin: Ethnomedizin—Medizinethnologie—medical anthropology. Reimer-Verlag, Berlin, pp 277–307
42. Campinha Bacote J (2019) Cultural competemility: a paradigm shift in the cultural competence versus cultural humility debate. Part I. Online J Issues Nurs 24(1). https://doi.org/10.3912/OJIN.Vol24No01PPT20
43. DasGupta S (2008) Narrative humility. Lancet 371(9617):980–981. https://doi.org/10.1016/S0140-6736(08)60440-7
44. von Uexküll T (1993) Rückmeldung als Modell interpersonaler Beziehungen: Psychosomatische Medizin als Beziehungsmedizin. Fundam Psychiatr 7:58–63
45. Brody H (1994) "My story is broken; can you help me fix it?" medical ethics and the joint construction of narrative. Lit Med 13(1):79–92. https://doi.org/10.1353/lm.2011.0169
46. World Medical Association (1983) Declaration of Geneva. World Medical Association, Geneva
47. Jaspers K (1963) General psychopathology. University of Chicago Press, Chicago
48. Achella S, Marazia C (2022) Lebensgeschichten. Karl Jaspers und die Humanistische Psychologie. Aufgang 18:67–83
49. Anzalone M (2023) Tra indeterminatezza e arbitrio: Jaspers e l'"arte" di scrivere Krankengeschichten. Studi Jaspersiani X(2022):13–30
50. Loriga S (2010) Le Petit x: De la biographie à l'histoire. Seuil, Paris, pp 112–113
51. Droysen JG (1893) Outline of the principles of history (Grundriss der Historik). Ginn, Boston
52. Pierosara S (2025) Narrative autonomy to the test of illness. In: Cenedese ML, Nicastro C (eds) Violence, care, cure: self perceptions within the medical encounter. Routledge, London, pp 151–162
53. Da Rocha AC (2014) Narrative autonomy: three literary models of healthcare in the end of life. Camb Q Healthc Ethics 23(2):200–208. https://doi.org/10.1017/S096318011300073X
54. Altschuler S, Huntington I, Antoniak M, Klein LF (2024) Clinician as editor: notes in the era of AI scribes. Lancet 404(10468):2154–2155. https://doi.org/10.1016/S0140-6736(24)02568-6
55. van Dongen J, Paul H (2017) Introduction: epistemic virtues in the sciences and the humanities. In: van Dongen J, Paul H (eds) Epistemic virtues in the sciences and the humanities. Springer, Cham, pp 1–14. https://doi.org/10.1007/978-3-319-48893-6_1
56. Adibi A, Cao X, Ji Z, Kaur JN, Chen W, Healey E, Jabbour S et al (2025) Recent advances, applications and open challenges in machine learning for health: reflections from research roundtables at ML4H 2024 symposium. arXiv:2502.06693

Chantal Marazia is assistant professor in History of Medicine and Bioethics, Università degli Studi di Milano. Her main research fields are the history of psychiatry and neurology, the history of bioethics and the history of human rights. She is editor of the psychopathological writings of Karl Jaspers.

Vasilija Rolfes is currently research associate at the Department of the History, Philosophy and Ethics of Medicine, Heinrich-Heine Universität Düsseldorf. Her research is focused on three primary areas: medical ethics, the stigmatization and discrimination of patients within the medical context, and the ethical implications of new medical technologies.

Fabio De Sio is research associate at the Department of the History, Philosophy and Ethics of Medicine, Heinrich-Heine Universität Düsseldorf. His research focuses on the institutional history of biology, the history and ethics of animal models in biomedicine, research ethics, the history of physiology, and the history of anthropology in the second half of the twentieth century.

Humanizing Machines: Unpacking the Care Deficit in Anthropomorphized Medical AI

13

Chetan Mahadev Shirvankar

Abstract

This chapter explores how data collected by embodied artificial intelligence (AI) contributes to potential care deficits in patient interactions. While the principles of narrative medicine highlight the importance of understanding the patient's personal story to deliver holistic care, there is a growing concern that the emphasis on data and metrics overshadows the essential human elements of empathy and personalized treatment. This chapter reviews Latour's Actor-Network Theory (ANT) which frames AI as an active participant in the healthcare network, influencing and reshaping the dynamics of the doctor-patient relationship. Additionally, Foucault's concept of biopolitics identifies these systems as tools of biopolitical control, managing and regulating patient care in ways that could reinforce existing power dynamics within the medical field. The tendency to anthropomorphize AI creates a misleading perception of objectivity or empathy, resulting in patients relying more on AI-driven recommendations, potentially at the expense of voicing their own preferences or concerns. This epistemic injustice could skew the shared decision-making process, making it less collaborative and more data dominated. The chapter concludes by addressing the "Control Problem," which concerns how to best align embodied AI with human goals and examines the modern laws of robotics being proposed to ensure that these systems in healthcare act in ways that enhance, rather than detract from, human-centered care.

Keywords

Artificial intelligence · Human-robot interaction · Medical robotics · Primary care · Digital health

C. M. Shirvankar (✉)
McGill University, Montreal, Quebec, Canada
e-mail: chetan.shirvankar@mail.mcgill.ca

M. Anzalone et al. (eds.), *Reconfiguring Human Autonomy*, Integrated Science 40,
https://doi.org/10.1007/978-3-032-14537-6_13

1 The Minute That Passes

It's that unsettling tension. When you're sure of what you want, but can't seem to choose. Trust yourself, or don't?. The question lingers, especially in situations where the decision isn't entirely yours to make. In clinical settings, decisions are rarely simple. Between the weight of data, the pressure of time, and the uncertainty of human lives, a physician learns to navigate doubt as part of the job. But what happens when the doubt disappears?

In a hospital room, artificial intelligence (AI) has become a quiet, confident presence. These systems do not hesitate, falter, or ask. They offer answers without flinching. They project certainty, even when they do not understand the question. And that certainty is alluring.

An AI system flags a sepsis risk. It predicts a heart failure readmission. It suggests antibiotics, recommends surgery, and estimates length of stay. These are not just possibilities. To the system, they are predictions with percentages and charts. To the clinician, they are nudges that can be difficult to ignore.

There is something deeply human about hesitation. The pause before a diagnosis is spoken aloud. The back-and-forth between two clinicians at odds over a patient's chart. The discomfort of not knowing and the courage it takes to sit with that uncertainty. This chapter is about that minute that it takes for a human to deliberate. The moment that allows for care, empathy, and intuition. As AI becomes embedded in clinical workflows, we risk losing that space. But we also have a chance to protect it.

We will explore how this pause is shaped and sometimes erased by systems of AI, drawing from fields like narrative medicine, actor-network theory, biopolitics, epistemic injustice, and the ethics of automation. These frameworks help us ask better questions: not just about how machines work, but about how we work with them. Not just what AI can do, but what it should do. This is not about rejecting technology. It is about designing systems that remember what it means to be human in the act of healing.

The minute may pass, but in that pause lies everything that matters.

2 When Machines Replace the Pause

Trust does not just appear in a vacuum. Medicine builds slowly through years of training, experience, and intuition. But in the age of AI, trust is being recalibrated. What once emerged from human interaction now comes from a glowing dashboard. Clinicians are increasingly being asked to rely on systems they did not build, cannot fully understand, and are often discouraged from questioning. These systems promise efficiency, consistency, and objectivity. And they deliver—but often at the cost of something harder to measure.

Consider this: a junior doctor enters a patient's room. The algorithm has already predicted a 79% chance of acute kidney injury. The recommendation is printed in red on the screen. There is no immediate urgency, but the system advises a nephrology consult. The doctor has not yet spoken to the patient. But the pause where they might have gathered their own impression is already shortened. Cognitive science offers some insight here. According to dual-process theory, we make decisions using two systems: a fast, intuitive one and a slower, more deliberative one [1, 2]. AI is essentially designed to work like the fast system, constantly feeding clinicians quick, confident answers. But the problem is not that these answers are fast. It is that they can replace the deliberative pause. This pause matters. It is where doubt lives, and doubt is not a flaw in medicine. It is a safeguard. It invites second opinions, deeper listening, and careful consideration. When AI removes that doubt, it does not just remove uncertainty. It eliminates an opportunity for reflection.

In practice, clinicians are already experiencing this shift. In survey after survey, they report feeling pressured to agree with algorithmic recommendations, not because they trust them entirely, but because to question them is to risk delay, inefficiency, or reprimand [3]. Over time, a pattern emerges: the system becomes a shortcut to decision-making. And the pause begins to vanish.

The promise of AI in healthcare has always been to support clinicians, not replace them. However, subtle design choices like how alerts are framed, how dashboards prioritize information, and how often peers accept recommendations all contribute to how much room clinicians feel they must think [4].

In one hospital study, simply changing the colour of an AI alert from red to yellow led to significantly more second opinions [5]. The red signal had implied urgency, even when none existed. A simple interface change gave clinicians the space to pause. This is the heart of the issue: not whether machines can be smart, but whether their design supports human thought. When machines replace the pause, they do not just offer answers, they change the questions we ask, and the way we ask them.

To protect the pause, we need to treat it as essential. It is not a luxury or a delay, but a critical part of care. And we need to design AI that respects it.

3 The Story in the Noise

Narrative medicine begins with a simple belief: that stories matter. Rita Charon, who pioneered the field, describes it as the art of recognizing, absorbing, interpreting, and being moved by the stories of illness [6]. It is a radical idea in a world of lab values and imaging reports that listening, truly listening, can itself be a form of diagnosis. But AI does not listen. It parses. It extracts. It quantifies. When a patient describes their pain, AI does not hear fear in the voice or notice hesitation before answering. It looks for structured input: duration, severity, keywords. If it can be measured, it is. Everything else is noise. Yet, in that noise, the story often hides.

Take the case of Mr. L., a 71-year-old with chronic obstructive pulmonary disease. He presents with shortness of breath. The AI, trained on thousands of similar profiles, flags a likely pneumonia and recommends admission. But the resident decides to spend a little more time with him. Mr. L. mentions his dog died 2 weeks ago. He has not been eating. Has barely left bed. The breathlessness, it turns out, is part grief, part dehydration, part anxiety. Not pneumonia. Not even close. When clinicians turn to dashboards before turning to patients, something is lost. Not just rapport, but diagnostic richness. Patients often offer subtle, layered narratives that do not fit neatly into electronic health record fields. A pause. A joke. A quiet contradiction. These are not statistical anomalies. They are clues.

In one study, medical students were asked to diagnose patients using either AI tools or traditional interviews [7]. Those using AI often arrived at faster answers. However, they also missed more atypical presentations, especially in patients with complex social histories. The stories had been filtered out. There is also a risk that clinicians begin to mimic the machines. As AI systems grow more prevalent, documentation starts to shift. Notes become templated. Language is standardized. The messy details like the smell of the room, the patient's word choice, and the moment they teared up often get edited out. And this matters. Storytelling is not just about the patient; it's about the clinician, too. Telling a story forces reflection. It transforms raw experience into meaning. Without it, care becomes transactional.

If AI systems are to be integrated meaningfully into clinical practice, they must make space for narrative. This could mean voice-enabled EHRs that preserve patient language, or interfaces that invite free-text alongside structured fields. More importantly, it means training clinicians to hold the story even when the system does not. Because stories do not slow down care, they ground it. And they remind us that we are treating not just a condition, but a person living through it.

4 Machines in the Network

Actor-Network Theory (ANT) challenges us to rethink how agency works in modern systems [8]. In ANT, people are not the only ones who act. Objects, technologies, protocols, and infrastructures also shape outcomes. It is not just about who holds power, but what forms it takes when machines, software, and humans work together. In healthcare, this theory is instrumental. A smart pump that controls medication doses does not just execute instructions. It influences workflow, changes nurses' interactions with patients, and reshapes accountability. Once AI enters this network, the dynamics shift again.

Take the example of a hospital where an AI tool is introduced to recommend treatment plans for cardiac patients. At first, it is presented as a second opinion. But over the months, its suggestions have become standard. Nurses begin documenting patient status in ways the system can interpret. Doctors stop offering alternative options during rounds unless the AI flags a problem. Eventually, a nurse who raises a concern based on intuition is told: "Just follow what the algorithm recommends." What began as a tool becomes an actor. And over time, an authority.

ANT helps explain this transformation. The AI is not just a passive assistant. It participates. It changes routines, expectations, and even language. It also reshapes roles: the nurse as executor, the doctor as confirmer, the AI as decision-leader. This co-production of action is not inherently bad. Technology has always shaped how medicine is practiced. But when roles shift without reflection, unintended consequences follow. The clinician may feel less responsible. The system may become less flexible. The patient may feel less heard. Moreover, machines in networks do not just shape practice. They shape policy. Hospital administrators might adopt AI to reduce costs or standardize care. Insurers may require its use to approve procedures. Medical boards might define competence in terms of how closely a doctor follows algorithmic recommendations. Suddenly, a machine that began as a helpful suggestion becomes a gatekeeper.

If we want to design systems that empower rather than displace, we must acknowledge the full extent of AI's influence. That means understanding its role not just at the bedside, but in the broader network of actors it connects: people, policies, tools, and spaces.

Only then can we ask the right question: not "What can the machine do?" but "What kind of actor do we want it to become?"

5 Biopolitics in the Waiting Room

Michel Foucault's idea of biopolitics describes how power operates not through force, but through the management of life itself [9]. It is not about coercing individuals, but about shaping populations, such as guiding behaviour, normalizing standards, and defining what it means to be healthy, productive, or compliant. In the context of AI in healthcare, biopolitics helps us see how digital tools do not just deliver care, they help define who is worthy of care, and on what terms [10].

Imagine a public health clinic that uses AI to manage patients with chronic conditions like diabetes and hypertension. The system sends reminders, tracks adherence, schedules follow-ups, and recommends diet changes based on wearable data. Patients who follow these recommendations are rewarded with shorter waiting times and streamlined care. Those who miss appointments or show inconsistent data patterns get flagged. Over time, the system begins to distinguish the "compliant" from the "noncompliant" not based on intent or need, but on algorithmic expectation.

No one explicitly punishes these patients. But care becomes conditional. Slower access, lesser attention, and subtle penalties are hidden in scheduling algorithms and triage protocols. This is power without visible enforcement, which indicates biopolitics at work.

AI does not need to declare a rule to enforce it. Its power lies in nudges, defaults, and silent exclusions. For instance, if the AI suggests that people who live alone are less likely to adhere to post-op protocols, those patients may be deemed higher risk and passed over for certain surgeries. If behavioural models show that people from lower-income neighbourhoods cancel more appointments, the system might allocate fewer resources to those clinics. These systems are not malicious. They are designed to maximize efficiency, reduce waste, and improve outcomes. But they operate on data, and data reflects the world as it is, with all its inequities baked in. If we do not question how these systems define success, we risk reinforcing the disparities we claim to address.

The concept of "soft control" is key here [11]. Unlike hard power, which is about rules and enforcement, soft control shapes choices by making some options easier than others. In digital healthcare, this might look like a mental health app that rewards daily check-ins, or a home-monitoring system that prioritizes patients who log their symptoms consistently. Over time, behaviour is shaped not by discussion or consent, but by design.

Biopolitical influence also shows up in population-level decisions. During the COVID-19 pandemic, predictive models were used to allocate ventilators, staff, and vaccines [12]. These models, while useful, also reflected value judgments: Which patients had the highest survival odds? Which communities were most "at risk"? Who counted as vulnerable? These were not just medical questions but ethical and political ones, too.

To resist harmful biopolitical effects, we must design AI systems that are aware of the biases they carry. This means including social determinants of health in data models, designing interventions that do not penalize the vulnerable, and building oversight mechanisms that account for how systems perform across different populations. We must also be honest about what we are asking AI to do. If we treat it as a neutral advisor, we miss its role as a subtle policymaker. The dashboard is not just a mirror of the world. It is a map. And every map simplifies, distorts, and directs. AI does not just help us deliver care. It helps define what care is and who deserves it. That is a responsibility we cannot leave to code alone.

6 Silenced Knowledge

Miranda Fricker's concept of epistemic injustice offers a powerful lens to examine how AI systems may unintentionally suppress certain kinds of knowledge in healthcare [13]. Epistemic injustice occurs when someone's knowledge, which is rooted in lived experience, is discredited, dismissed, or ignored. In medicine, this has long been an issue for marginalized groups. With AI, the stakes become even higher.

Consider the mother of a child who insists that something is not right. The algorithm disagrees. The clinicians trust the AI, which has reviewed thousands of similar cases. But several hours later, the child begins seizing. It turns out the mother was right, her child had a rare condition that the system did not recognize. Her voice was not just overlooked. It was excluded by design. This is not new. Women, racialized patients, people with disabilities, and those with chronic or poorly understood conditions have often struggled to be believed in clinical settings. What AI risks doing is making that disbelief more systematic. An algorithm trained on data that underrepresents these voices will not just inherit that bias. It may magnify it.

There is a deep irony in a system that claims to be "evidence-based" while routinely sidelining certain types of evidence. Pain, fatigue, and mental health symptoms often do not show up clearly in diagnostic tests. They require listening, context, and trust. These are precisely the qualities that many AI systems are not designed to accommodate.

We must also consider how epistemic injustice plays out in the design of AI itself. Who gets to define what data is relevant? Who decides how it is labelled, or what counts as a successful outcome? Often, the patients most affected by AI-driven decisions are less likely to have been involved in their creation. Take, for example, a predictive tool used in emergency departments to determine which patients will likely return within 72 h. Suppose the model is trained mostly on urban, insured, English-speaking patients. In that case, it may fail to account for the lived realities of others, like non-English speakers, people experiencing homelessness, or undocumented immigrants. When these individuals are misclassified, they do not just receive suboptimal care. They are further excluded from the system's understanding of what a "typical" patient looks like.

Epistemic injustice also affects healthcare professionals. Junior staff, nurses, or community health workers may offer insights that do not align with the AI's prediction, only to be dismissed. Over time, these professionals may stop speaking up, leading to a loss of human knowledge that cannot be easily captured by data. The solution is not just more data. It is better data with a more inclusive approach to what counts as knowledge. This includes:

- Designing AI systems that allow for narrative input alongside quantitative data.
- Creating feedback loops where patients and clinicians can challenge or refine algorithmic recommendations.
- Including diverse stakeholders in the development and validation of models.
- Auditing outcomes not just for accuracy but for fairness across demographics.

We must move beyond a narrow view of intelligence as pattern recognition. True clinical insight often comes from anomalies. The case that does not fit, the symptom that should not be there, the story that disrupts the trend. Listening to these stories, even when they contradict the algorithm, is not inefficient. It is essential. Because the cost of ignoring them is not just a diagnostic error, it is the erosion of trust and the silencing of voices that were never meant to be excluded in the first place.

7 The Illusion of Control

One of the most seductive features of AI in healthcare is the appearance of control. Algorithms provide crisp outputs, confidence scores, and ranked recommendations. To clinicians, especially those overwhelmed by caseloads or administrative demands, these outputs offer clarity or at least the illusion of it. But control is not the same as certainty. And in medicine, the line between the two can blur quickly.

Guglielmo Tamburrini describes the gap between algorithmic decision and robotic action as one of speed [14]. A system that acts before a human has time to intervene. In critical care, this gap can feel even sharper. AI may recommend treatment adjustments, predict deterioration, or escalate alerts without waiting for context. These recommendations often arrive wrapped in urgency, colour-coded, timestamped, and presented as best practice.

But what if the system is wrong?

Consider the case of a trauma surgeon reviewing a triage algorithm's assessment. The AI classifies a patient as stable. Vitals are within range. But something feels off. The patient is too quiet, too still. The surgeon trusts her gut, orders a scan, and finds internal bleeding. The AI missed it not because it malfunctioned, but because it followed rules that the patient did not fit. These moments matter. Because what is presented as objective is often incomplete. AI systems are trained on historical data, and they learn to optimize for patterns. But patterns are not people. They do not account for intuition, cultural nuance, or the emotional intelligence that skilled clinicians bring to their work.

There is also the danger of false reassurance. When an AI gives a diagnosis with 92% confidence, what does that really mean? To many, it feels definitive, even if the remaining 8% represents real, serious uncertainty. Over time, clinicians may become less likely to question high-confidence outputs, even when their instincts suggest something else.

In aviation, pilots are trained to challenge automated systems, not defer to them blindly. In healthcare, we need a similar mindset. The ability to override, question, and pause should not be treated as an inconvenience. It should be built into the design.

We must also be wary of the psychological effects of AI reliability. A system that is right 95% of the time can build a kind of dependence. But when it fails, when it misses that rare disease, or misclassifies a stroke. The consequences are immediate and devastating. The more we rely on AI, the harder it becomes to re-engage our own uncertainty when we need it most. The illusion of control is powerful because it feels efficient. But efficiency without room for doubt can become dangerous. The best clinicians are not the fastest or most confident; they are the ones who know when to slow down, when to ask again, when to listen to a hunch. AI should not replace that capacity. It should support it. And to do so, it must be transparent not just about what it knows but also what it does not. Because in medicine, real control begins with humility, the recognition that not everything that matters can be measured, and not every answer should be automated.

8 Autonomy Without Accountability

Artificial intelligence in healthcare is increasingly described as "autonomous", capable of learning, adapting, and making decisions without direct human instruction. But this autonomy raises a difficult question: if a machine can act on its own, who is responsible for what it does?

AI systems make choices in search of an author. They execute decisions based on programmed logic and learned data patterns, but without moral reasoning or intent. The system may choose a course of action but cannot explain why in human terms. It cannot reflect on the ethical implications or justify itself when something goes wrong.

This gap between autonomy and accountability creates a dangerous blind spot in healthcare. In the past, if a clinician made a poor decision, there was a clear line of responsibility. As AI becomes more embedded in diagnostic and treatment workflows, those lines blur. Was the doctor wrong to follow the algorithm? Or to override it? Was the system at fault? Or the data it was trained on?

Legal scholars and ethicists are already wrestling with this challenge. If a surgical robot causes harm because of a software glitch, who is liable? Is it the manufacturer, the hospital, the technician, or the surgeon who pressed the start button? In many jurisdictions, existing laws struggle to keep pace with these scenarios.

Beyond liability, there's the issue of moral accountability. Patients want and deserve an explanation when something goes wrong. But how do we respond if the decision was made by an AI model trained on millions of data points, and no one can fully explain why it reached that conclusion? There is also the risk of moral disengagement among clinicians. When decisions come prepackaged from a machine, it becomes easier to shift responsibility. "The system said so" is not just a technical explanation but can become a shield against self-reflection. Over time, this erodes the clinician's role as a moral agent.

Transparency is often offered as a solution. Explainable AI (XAI) aims to make algorithms more interpretable, showing features that influence a decision or explain why a specific recommendation was made [15]. But explanations are not always sufficient. Understanding how a decision was reached is not the same as justifying it ethically.

To bridge the gap, we need systems that are not only explainable but embedded in ethical frameworks. This means:

- Designing AI tools that prompt human oversight, not replace it.
- Creating shared accountability models between developers, clinicians, and institutions.
- Establishing clear protocols for when and how to override algorithmic decisions.
- Including ethical deliberation in AI governance — not just technical performance metrics.

Autonomy in machines should never come at the expense of responsibility in people. AI can assist in decision-making but cannot bear the moral weight of those decisions. That weight remains and must remain with us. In healthcare, every decision touches a life. And no machine can take full responsibility for that, no matter how advanced.

9 Designing for Doubt

If there is one theme that has emerged across the previous sections, it is this: doubt is not the enemy of healthcare; instead, it is a vital part of it. The instinct to hesitate, ask again, and seek context before acting is not an inefficiency. They are safeguards. And yet, many current AI systems are designed to minimize or bypass them. What would it mean to create AI systems that protect, rather than suppress, human doubt?

First, we must reconsider the interface. Most clinical AI tools are built for speed and clarity. Their design emphasizes decisiveness—red alerts, ranked recommendations, ticking clocks. But this aesthetic of certainty can be misleading. Interfaces should be designed not only to inform but also to invite reflection. That means showing uncertainty ranges, offering alternate options, and allowing space for clinicians to annotate or question. Take, for instance, a decision support tool for stroke triage. Rather than providing a single course of action, it could present probabilities for multiple diagnoses, prompt the clinician to consider rare but dangerous alternatives, or link to recent similar cases. These nudges would not slow care; they would strengthen it.

Second, systems must encourage feedback loops. Clinicians should be able to report when an AI's recommendation was inaccurate or misleading, and those reports should feed back into model refinement. This improves performance and builds trust as the system is not static, but responsive.

Third, doubt should be built into the workflow. Some hospitals require a second opinion before high-risk procedures, regardless of AI recommendation. Others use interdisciplinary rounds to discuss complex cases as a team. These practices predate AI but become even more important when algorithms are involved. Decision-making should remain collaborative and not outsourced.

Fourth, education [16] is critical. Medical training must include AI literacy, not just how to use tools, but how to question them. Clinicians should learn when to trust the system and when to trust themselves. This requires teaching technical skills, ethical reasoning, and situational awareness.

There is also a role for policy. Institutions should not penalize clinicians for deviating from AI recommendations, provided they document their reasoning. Legal frameworks should support professional judgment, not discourage it.

Finally, we must resist the temptation of framing AI as a human care replacement. Instead, we should design systems that extend our humanity. That might mean integrating patient stories into the algorithm's inputs. It might mean offering explanations in plain language so patients can understand and challenge decisions. It might mean building tools that are not just efficient, but empathetic.

Doubt is not a flaw. It is a feature of reasonable care. And in the rush to automate, we cannot afford to lose it. Designing for doubt means designing for humility. For curiosity. For the pause that allows us to ask, "What else might be true?". In that pause, there is space to reconsider, listen, and see what the machine may have missed. And that space, which is inherently fragile, fleeting, and profoundly human, is where real healing begins.

10 The Human Interval

Machines act. Humans hesitate. And in that hesitation lies the possibility of care, not just technical competence, but compassion, intuition, and moral judgement. As AI becomes more capable and embedded in healthcare, the need to protect the space for that hesitation becomes even more urgent. This space—the human interval—is where the machine's certainty meets the clinician's doubt, where data meets narrative. Where efficiency meets empathy, it is not a delay to be eliminated, but a moment to be honoured.

In practice, the human interval might look like a nurse pausing before administering medication to double-check a dosage that "doesn't feel right." Or a physician sitting a few minutes longer with a patient whose body language tells a different story than the scan. Or a team discussing a plan one more time, even though the algorithm gave a green light.

These moments do not slow care. They deepen it.

This is where the Extended Mind Thesis becomes especially relevant [17]. If cognition is not confined to the brain but stretches into our tools, environments, and relationships, then AI systems are not just external aids; they become part of how decisions are made. In healthcare, this distributed cognition adds power, but also risk. It underscores the need to ensure that machines do not displace judgement but expand it. The human interval is not simply a pause. It is a shared cognitive space where human insight and machine precision meet, where ethics, empathy, and narrative take shape alongside data.

To safeguard this space, it is worth remembering the roots of machine ethics. Isaac Asimov's original laws of robotics imagined a future where machines must never harm a human, obey humans unless it causes harm, and protect themselves unless those conflicts with the first two laws [18]. While these were fictional, they sparked real-world debates about responsibility and control. Today's challenges are more complex, and modern laws of robotics build on those ideas by emphasizing transparency, explainability, and alignment with human values [19]. These new principles call for AI systems to do more than avoid harm. They must support clinical reasoning, reinforce trust, and respect the relational fabric of care. When we prioritize speed above all else, we risk losing something essential, not just the richness of human judgement, but the sense of connection that sustains both patients and clinicians. Medicine is not just a science of the body. It is a practice of presence. This does not mean rejecting AI. It means demanding more of it, not just in terms of accuracy, but in how it supports the values that matter most. Systems should be built not just to inform decisions, but to respect them. Not just to produce outcomes, but to foster understanding.

Ultimately, AI should not diminish the clinician's role but expand their capacity to care. It should make space for second thoughts, questions, and relationships. It should serve as a partner, not a master. The human interval is where a patient's story can be heard fully. Where a diagnosis becomes more than a label. Where care becomes more than treatment.

As we continue to integrate machines into our health systems, let's not forget that what matters most may not be what the system gets right but what it might miss, and who is there to notice. The minute that passes. The breath that catches. The silence that allows for reconsideration.

That is where healing begins. Let's make sure we leave room for it.

Competing Interests The author has no conflicts of interest to declare that are relevant to the content of this chapter.

References

1. Tversky A, Kahneman D (1974) Judgment under uncertainty: heuristics and biases. Science 185(4157):1124–1131. https://doi.org/10.1126/science.185.4157.1124
2. Evans JS, Stanovich KE (2013) Dual-process theories of higher cognition: advancing the debate. Perspect Psychol Sci 8(3):223–241. https://doi.org/10.1177/1745691612460685
3. Sauerbrei A, Kerasidou A, Lucivero F, Hallowell N (2023) The impact of artificial intelligence on the person-centred, doctor-patient relationship: some problems and solutions. BMC Med Inform Decis Mak 23(1):73
4. Miller K, Mosby D, Capan M, Kowalski R, Ratwani R, Noaiseh Y et al (2018) Interface, information, interaction: a narrative review of design and functional requirements for clinical decision support. J Am Med Inform Assoc 25(5):585–592
5. Cabitza F, Campagner A, Sconfienza L (2020) As if sand were stone. New concepts and metrics to probe the ground on which to build trustable AI. BMC Med Inform Decis Mak 20:219. https://doi.org/10.1186/s12911-020-01224-9
6. Charon R (2007) What to do with stories: the sciences of narrative medicine. Can Fam Physician 53(8):1265–1267
7. Tu T, Schaekermann M, Palepu A, Saab K, Freyberg J, Tanno R et al (2025) Towards conversational diagnostic artificial intelligence. Nature 642:442–450. https://doi.org/10.1038/s41586-025-08866-7
8. Latour B (1987) Science in action: how to follow scientists and engineers through society. Harvard University Press
9. Miller T (2010) Michel Foucault, the birth of biopolitics: lectures at the Collège de France, 1978–79. Int J Cult Policy 16(1):56–57
10. Wrangel C (2024) The biopolitics of algorithmic governmentality: how the US military imagines war in the age of neurobiology and artificial intelligence. Secur Dialogue 55(4):349–367. https://doi.org/10.1177/09670106241226622
11. Ghosh A, Saini A, Barad H (2025) Artificial intelligence in governance: recent trends, risks, challenges, innovative frameworks and future directions. AI Soc 40:1–23
12. Junger N, Hirsch O (2024) Ethics of nudging in the COVID-19 crisis and the necessary return to the principles of shared decision making: a critical review. Cureus 16(4):e57960. https://doi.org/10.7759/cureus.57960
13. Fricker M (2007) Epistemic injustice: power and the ethics of knowing. Oxford University Press
14. Amoroso D, Tamburrini G (2020) Autonomous weapons systems and meaningful human control: ethical and legal issues. Curr Robot Rep 1(4):187–194. https://doi.org/10.1007/s43154-020-00024-3
15. Confalonieri R, Coba L, Wagner B, Besold T (2020) A historical perspective of explainable artificial intelligence. WIREs Data Min Knowl Discov 11:e1391. https://doi.org/10.1002/widm.1391
16. Tolentino R, Baradaran A, Gore G, Pluye P, Abbasgholizadeh-Rahimi S (2024) Curriculum frameworks and educational programs in AI for medical students, residents, and practicing physicians: scoping review. JMIR Med Educ 10(1):e54793
17. Clark A, Chalmers D (1998) The extended mind. Analysis 58(1):7–19
18. Asimov I (2004) I, Robot. Spectra
19. Spiekermann S (2021) What to expect from IEEE 7000: the first standard for building ethical systems. IEEE Technol Soc Mag 40(3):99–100

Chetan Mahadev Shirvankar is a physician scientist at McGill University and Mila Quebec AI Institute. Trained as a clinician, he works at the intersection of artificial intelligence, robotics, and primary care, building tools that put people at the centre of healthcare innovation. He founded the Human Robot Project to explore how robots can support elderly care in ethical, meaningful ways. His work has been shared at international conferences, published in peer-reviewed journals, and recognized through his role on the editorial board of Biomedical Engineering Online. Through both research and advocacy, he is driven by one question: how can we design technology that respects, rather than replaces, human autonomy?

MIX
Papier aus verantwortungsvollen Quellen
Paper from responsible sources
FSC® C105338

If you have any concerns about our products,
you can contact us on
ProductSafety@springernature.com

In case Publisher is established outside the EU,
the EU authorized representative is:
Springer Nature Customer Service Center GmbH
Europaplatz 3, 69115 Heidelberg, Germany

Printed by Libri Plureos GmbH
in Hamburg, Germany